Srdjan Glisic

Design of Fully Integrated 60GHz OFDM Transmitter

Srdjan Glisic

Design of Fully Integrated 60GHz OFDM Transmitter

Design of fully integrated mm-wave PAs, passive filters, PLLs in SiGe BiCMOS technology

Südwestdeutscher Verlag für Hochschulschriften

Impressum/Imprint (nur für Deutschland/only for Germany)
Bibliografische Information der Deutschen Nationalbibliothek: Die Deutsche Nationalbibliothek verzeichnet diese Publikation in der Deutschen Nationalbibliografie; detaillierte bibliografische Daten sind im Internet über http://dnb.d-nb.de abrufbar.
Alle in diesem Buch genannten Marken und Produktnamen unterliegen warenzeichen-, marken- oder patentrechtlichem Schutz bzw. sind Warenzeichen oder eingetragene Warenzeichen der jeweiligen Inhaber. Die Wiedergabe von Marken, Produktnamen, Gebrauchsnamen, Handelsnamen, Warenbezeichnungen u.s.w. in diesem Werk berechtigt auch ohne besondere Kennzeichnung nicht zu der Annahme, dass solche Namen im Sinne der Warenzeichen- und Markenschutzgesetzgebung als frei zu betrachten wären und daher von jedermann benutzt werden dürften.

Verlag: Südwestdeutscher Verlag für Hochschulschriften GmbH & Co. KG
Heinrich-Böcking-Str. 6-8, 66121 Saarbrücken, Deutschland
Telefon +49 681 37 20 271-1, Telefax +49 681 37 20 271-0
Email: info@svh-verlag.de

Approved by: Cottbus, BTU, Diss., 2010

Herstellung in Deutschland:
Schaltungsdienst Lange o.H.G., Berlin
Books on Demand GmbH, Norderstedt
Reha GmbH, Saarbrücken
Amazon Distribution GmbH, Leipzig
ISBN: 978-3-8381-2917-4

Imprint (only for USA, GB)
Bibliographic information published by the Deutsche Nationalbibliothek: The Deutsche Nationalbibliothek lists this publication in the Deutsche Nationalbibliografie; detailed bibliographic data are available in the Internet at http://dnb.d-nb.de.
Any brand names and product names mentioned in this book are subject to trademark, brand or patent protection and are trademarks or registered trademarks of their respective holders. The use of brand names, product names, common names, trade names, product descriptions etc. even without a particular marking in this works is in no way to be construed to mean that such names may be regarded as unrestricted in respect of trademark and brand protection legislation and could thus be used by anyone.

Publisher: Südwestdeutscher Verlag für Hochschulschriften GmbH & Co. KG
Heinrich-Böcking-Str. 6-8, 66121 Saarbrücken, Germany
Phone +49 681 37 20 271-1, Fax +49 681 37 20 271-0
Email: info@svh-verlag.de

Printed in the U.S.A.
Printed in the U.K. by (see last page)
ISBN: 978-3-8381-2917-4

Copyright © 2011 by the author and Südwestdeutscher Verlag für Hochschulschriften GmbH & Co. KG and licensors
All rights reserved. Saarbrücken 2011

Acknowledgements

While working in the IHP's Circuit Design department I had support from many colleagues. At this place I want to thank those who helped me with the thesis.

First of all, I want to thank to my supervisor Prof. Rolf Kraemer who offered me the opportunity to work at IHP and supported my work ever since.

I especially want to thank Dr. J. Christoph Scheytt who supported my work and was always available for many useful discussions.

I wish to thank all my colleagues from the project and department for their help, useful discussions, support and friendship. I thank Dr. Wolfgang Winkler, Dr. Frank Herzel, Dr. Eckhard Graß, Mohamed Elkhouly, Dr. Prabir Datta, Dr. Yaoming Sun, Dr. Miloš Krstić, Miroslav Marinković, Johannes Borngräber, Maxim Piz, Dr. Chang-Soon Choi, Dr. Klaus Schmalz, Dr. Gerhard Fischer, Falk Korndörfer, Markus Ehrig, Markus Petri and Frank Popiela.

Želim na ovom mjestu da se zahvalim za svu ljubav i podršku koju sam imao od moje majke i brata, a naročito za njihovu podršku za vrijeme mojih studija.

Finally, my love and gratitude go to my wife Izabela for her great help, support and love. Writing this thesis was possible thanks to her constant support. My thanks go to my son Adam, who brings me joy every day.

Contents

Abstract vii
Zusammenfassung ix
Chapter 1.   Introduction 1
1.1.   Wireless Communication at 60 GHz 1
1.2.   IHP's SiGe:C BiCMOS Technology 4
1.3.   Thesis Objective and Organization 5
Chapter 2.   Transmitter Achitecture 7
2.1.   Introduction 7
2.2.   Application Scenarios and TX Requirements 7
   2.2.1.   Link Budget Calculation 8
2.3.   TX Topology 9
   2.3.1.   Version I TX Topology 10
   2.3.2.   Version II TX Topology 10
   2.3.3.   Version III TX Topology 10
   2.3.4.   Upconversion Mixer 11
   2.3.5.   Preamplifier 12
2.4.   Summary 13
Chapter 3.   Phase–locked Loop 15
3.1.   Introduction 15
3.2.   PLL Theory 16
   3.2.1.   Type I PLL 16
      3.2.1.1.   Type I PLL Components 16
      3.2.1.2.   Type I PLL Operating Principle 18
   3.2.2.   Type II PLL 20
      3.2.2.1.   Type II PLL Components 20
      3.2.2.2.   Type II PLL Linear Model 22
      3.2.2.3.   Type II PLL Stability Analysis 25
   3.2.3.   PLL Phase Noise Properties 27
      3.2.3.1.   Phase Noise of the Input Signal 27
      3.2.3.2.   VCO Phase Noise 28
      3.2.3.3.   LPF Noise 29
3.3.   RMS Phase Error Optimization 30
3.4.   Calculation of PLL Parameters 31

3.4.1. Parameter Calculation Recipe for a Third Order PLL 31
3.4.2. Parameter Calculation Recipe for a Fourth Order PLL 32
3.4.3. A New Parameter Calculation Recipe for a Fourth Order PLL 36
3.5. Comparison of Different PLL Topologies 41
3.5.1. Dual Loop PLL 41
3.5.2. Comparison of a Third, Forth Order and Dual Loop PLLs 42
3.6. PLL for AFE version I 44
3.7. PLL for AFE version II 47
3.8. Summary 47

Chapter 4.   Image–rejection Filter 49
4.1. Introduction 49
4.2. Filter Design Theory 49
4.2.1. Two–Port Network Characterization 50
4.2.2. Filter Parameters 51
4.2.3. Typical Filter Response Approximations 52
4.2.4. Lowpass Prototype Filters 53
4.2.5. Element and Frequency Transformation 55
4.2.6. Immittance Inverters 57
4.2.7. Filters with Immittance Inverters 59
4.2.8. Richards' Transformation 61
4.2.9. End–Coupled Microstrip Filters 62
4.2.10. Parallel–Coupled Microstrip Filters 65
4.2.11. Hairpin Microstrip Filters 66
4.2.12. Selective Microstrip Filters with Transmission Zeros 69
4.3. Filter Losses 71
4.3.1. Effects of Filter Losses on the Lowpass Frequency Response 72
4.3.2. Effects of Filter Losses on the Bandpass Frequency Response 72
4.3.3. Microstrip Filter Losses 74
4.4. Design of the 60 GHz Image–Rejection Filter 75
4.4.1. Specifications of the Image–rejection Filter 75
4.4.2. The State–of–the–Art in 60 GHz Filters 75
4.4.3. Filter Simulation and Substrate Definition 76
4.4.4. Filter Design 77
4.4.4.1. Broadband Hairpin Filter 79
4.4.4.2. Broadband Parallel–Coupled Asymmetrically Tuned Filter 80
4.4.4.3. Parallel–Coupled Asymmetrically Tuned Filter for the IEEE 802.15.3c Standard . 82
4.4.4.4. Lumped Element Filter for the IEEE 802.15.3c Standard 83

	4.4.4.5.	Narrowband on–Board Filter	85
4.5.	Summary		86

Chapter 5. Power Amplifier 89

- 5.1. Introduction 89
- 5.2. Power Amplifier Theory 90
  - 5.2.1. Stability Considerations of an Amplifier 90
  - 5.2.2. Small-Signal Operation of an Amplifier 93
  - 5.2.3. Amplifier Linearity 94
    - 5.2.3.1. AM-AM Distortion and 1dB Compression Point 95
    - 5.2.3.2. AM-PM Distortion 96
    - 5.2.3.3. Intermodulation Distortion 97
  - 5.2.4. Linearity of Cascaded Amplifiers 98
  - 5.2.5. Efficiency and PA Classes 99
  - 5.2.6. Load-Pull Measurements 100
  - 5.2.7. Power Combining Techniques 101
- 5.3. Modeling of Distributed Passive Elements for Matching 102
- 5.4. Power Amplifier Design in the 60 GHz Range 105
  - 5.4.1. PA System Requirements 105
  - 5.4.2. PA Schematic Design 105
  - 5.4.3. PA Layout and Post-layout Simulation 110
- 5.5. Power Amplifier Measurement 113
  - 5.5.1. Measurement Setup 113
  - 5.5.2. Measurement Results 115
  - 5.5.3. Comparison with the State-of-the-Art 120
- 5.6. Summary 121

Chapter 6. Transmitter Integration 123

- 6.1. Introduction 123
- 6.2. TX Integration and Board Design 123
- 6.3. Measurement Results 124
  - 6.3.1. TX Version I Measurement Results 125
  - 6.3.2. TX Version II Measurement Results 127
  - 6.3.3. TX Version III Measurement Results 129
  - 6.3.4. Link-Budget Calculation for the Version III AFE 132
- 6.4. Comparison with the State-of-the-Art 133
- 6.5. Summary 134

Chapter 7. Conclusions 137

References 141

List of Figures 147
List of Tables 153
Publications 155
List of Acronyms and Symbols 159

Abstract

The goal of this thesis is the analysis of the challenges and finding solutions for the design of mm-wave transceivers. The work presented here is focused on design of transmitter (TX) components, which are critical for the performance of the whole analog front-end. Phase-locked loop (PLL) phase noise is optimized, an image-rejection filter and a high 1 dB compression point (P1dB) power amplifier (PA) are designed.

The PLL phase noise optimization is presented and different PLL topologies are compared. A new optimized recipe for calculating PLL parameters of a forth order PLL is presented. Using this approach the spurious sidebands can be reduced by up to 10 dB.

The image-rejection filter chapter analyzes the challenges related to the design of the integrated image–rejection filter. The analysis presented here is the first on integrated filters for the 60 GHz band, because the previously published work dealt with on-board filters. The main problems related to the design of integrated filters arise from the low quality factor of the integrated resonators. The effects are high insertion loss and low selectivity. Two measures to reduce the insertion loss of the image–rejection filters were suggested. One is to design the filter as broadband. This measure deteriorates selectivity, so the minimum required image–rejection will limit the width of the passband. The second measure is to design the filter as broadband with non-equidistant transmission zeros (i.e. asynchronously tuned filter). This measure will improve both the insertion loss and the image–rejection.

The challenges related to the design of mm-wave PAs with high P1dB are analyzed and the procedure of the PA design is presented. The difficulties related to the PA design and layout are discussed and optimum solutions presented. Limits of different power combining techniques for integrated PAs are discussed. Effects of poor on-chip ground connection are analyzed. Different causes for P1dB degradation are analyzed. The produced PA features a differential cascode topology. The layout is symmetrical and presents a virtual ground on the symmetry line for the differential signal. The optimized schematic and a symmetrically drawn layout resulted in a 17 dBm measured P1dB. It was the highest reported P1dB in 60 GHz SiGe PAs when it was published.

The fully integrated TX was used for data transmission with data rate of 3.6 Gbit/s (with coding 4.8 Gbit/s) over 15 meters. This is the best result in the class of 60 GHz AFEs without beamforming.

Zusammenfassung

Das Ziel dieser Arbeit ist die Analyse der Anforderung an mm-Wellen Funksysteme und das Aufzeigen von Lösungswegen zum erfolgreichen Entwurf solcher integrierten Systeme in SiGe Technologie. In der hier vorgelegten Arbeit untersuchte ich schwerpunktmäßig den Entwurf von Sender Komponenten, welche entscheidend für die Leistungsfähigkeit des gesamten HF-Funksystems sind. Das Phasenrauschen der Phasenregelschleife (PLL) wurde optimiert, Spiegelfrequemzfilter und Leistungsverstärker mit hohem 1 dB Kompressions-Punkt (P1dB) wurden entworfen und als integriertes Senderchip hergestellt.

Es werden hier Verfehren zur Optimierung des PLL Phasenrauschens dargestellt und verschiedene PLL Topologien verglichen. Ein neues optimiertes Verfahren zur Berechnung der Parameter einer PLL vierter Ordnung wird vorgestellt. Die Anwendung dieses Verfahrens ermöglichte es, die unerwünschten PLL Seitenbänder um bis 10 dB zu verringen.

Im Kapitel Spiegelfrequenzfilter werden die Herausforderungen an den Entwurf von integrierten Spiegelfrequenzfiltern analysiert. Die hier dargestellte Analyse ist die erste für integrierte Filter im 60 GHz Frequenzbereich. Bisherige Veröffentlichungen basierten auf Dünnfilmtechnologiefilter. Die Hauptprobleme bezüglich des Entwurfs integrierter Filter basieren auf dem geringen Gütefaktor der integrierbaren Resonatoren. Die Folgen sind hohe Durchgangsdämpfung und geringe Selektivität. Zwei Methoden zur Verringerung der Durchgangsdämpfung werden dargestellt. Eine besteht im Entwurf dieser Filter als Breitbandfilter. Diese Methode verringert die Selektivität, so dass die minimal geforderte Unterdrückung der Spiegelfrequenz die Bandbreite begrenzt. Die zweite Methode besteht darin, den Filter als Breitband mit nicht-äquidistanten Transmissions-Nullstellen zu entwerfen. Diese Methode verbessert gleichzeitig die Durchgangsdämpfung und die Unterdrückung der Spiegelfrequenz.

Die Anforderungen an den Entwurf von mm-Wellen Leistungsverstärkern mit hohem P1dB werden analysiert und ein Entwurfsverfahren für einen Leistungsverstärker wird dargelegt. Die Schwierigkeiten bezüglich des Designs und Layoutentwurfs des Leistungsverstärkers werden diskutiert und optimierte Lösungswege werden dargestellt. Die Grenzen verschiedener Richtkopplungstechniken für integrierte Leistungsverstärker werden präsentiert. Die Effekte schwacher Masseverbindungen auf dem Chip werden analysiert. Verschiedene Ursachen für P1dB Degradation werden untersucht. Der hergestellte Leistungsverstärker hat eine differentielle Kaskode Topologie. Sein Layout ist symmetrisch und besitzt eine virtuelle Masse an der Symmetrielinie für das differentielle Signal. Das optimizierte Schaltbild und das symmetrisch entworfene Layout führten zu einem gemessenen P1dB von 17 dBm. Das war der höchste bisher veröffentlichte P1dB bei 60 GHz für einen Leistungsverstärker in SiGe Technologie, als die Meßergebnisse publiziert wurden.

Der vollintegrierter Sender wurde für eine Datenübertragung über 15 m mit einer Datenrate von 3,6 Gbit/s (mit Kodierung 4,8 Gbit/s) verwendet. Das ist das beste Ergebnis in der Klasse der 60 GHz Funksysteme ohne Beamforming.

Chapter 1

Introduction

1.1. Wireless Communication at 60 GHz

Ever since the begin of radio-frequency (RF) wireless communication some 100 years ago, the development of communication systems has been rapid. The need for larger volume and data rate has grown tremendously. The limited RF bandwidth has pushed the development of the communication systems in the direction of optical fiber-based systems. Today, most cities have good fiber connection for transmission of large amount of data, but there is no such broadband connection to the end user. This is due to a set of problems (costly installation, long lead time, est.) commonly referred to as the last-mile problem. Furthermore, the available technology is not suitable for high data-rate, small distance, low cost wireless communication.

There is a large ISM band of up to 9 GHz around 60 GHz that offers the possibility of wireless communication systems with large data-rate and has the capacity to offer the solution to the last-mile problem [1.1] – [1.3]. This frequency band had the attention already some years ago from different companies and research institutes [1.4], [1.5]. The uniqueness of the 60 GHz band is related to the atmospheric absorption. Figure 1. 1 shows the atmospheric absorption of the mm-wave frequencies, with an attenuation peak at 60 GHz due to oxygen. This may seem as a disadvantage, but for a short distance it plays no significant role in the systems link budget [1.2]. There are several reasons why the 60 GHz wireless communication is so attractive:

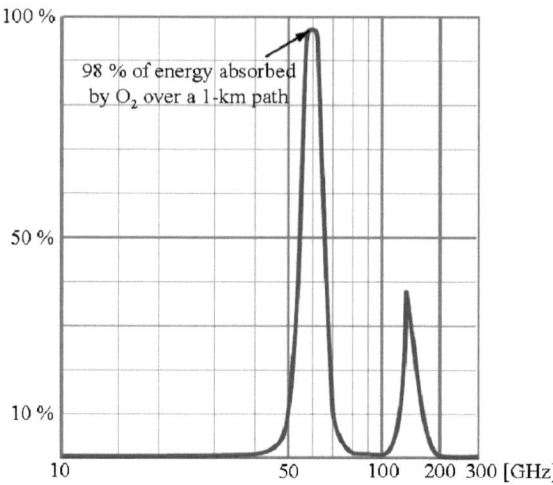

Figure 1.1 Atmospheric absorption of mm-wave frequencies over a 1-km path (from [1.3]).

1. The large available bandwidth offers data transmission in the order of several Gbit/s over the distance of several meters.

2. The unlicensed band means no need for time and money investment for companies to get the license.

3. The oxygen absorption enables frequency reuse at much smaller distances than at other mm-wave frequencies.

4. Antennas at 60 GHz are much more directive compared with the antennas at lower frequencies [1.2]. This together with the oxygen absorption makes a 60 GHz signal hard to intercept, because the signal will be focused and at larger distances too low for interception. This makes data transmission with such wireless systems much safer compared with the systems that operate at lower frequencies.

These reasons were the driving force in the development of a 60 GHz wireless communication system in the WIGWAM project [1.6] and the follow-up EASY-A project [1.7].

We should point out that the 60 GHz band has one important disadvantage. The free-space loss (L_{FS}) in this band is high. It is calculated using Friis transmission formula:

$$L_{FS}[dB] = 10\log\left(\left(\frac{4\pi df}{c}\right)^2\right) = 20\log(d[m]) + 20\log(f[GHz]) + 32.45 dB \quad (1.1)$$

where f is the signal frequency (in Hz), d is the distance between transmitter (TX) and receiver (RX) antennas (in m) and c is the velocity of EM waves in vacuum (in m/s). According to this formula the free-space loss at 60 GHz over 1 m is 68 dB. This is much higher loss than, for example, 46 dB at 5 GHz. This makes design of 60 GHz wireless communication systems much more difficult and limits the maximum distance of communication. In addition to this, the link budget is further strained by the limited TX output power and relatively high RX noise figure. For the line-of-sight (LOS) systems, high gain antennas can solve the link budget problem, and for the non line-of-sight (NLOS) systems, antenna arrays with beamforming can be used, but at the cost of higher complexity and price.

Figure 1. 2 shows the current status of the spectrum allocation in different countries. The unlicensed spectrum spreads from 57 to 66 GHz, and most countries have a common range from 59 to 64 GHz. Different standards apply regarding the maximum radiated output power and maximum equivalent isotropic radiated power (EIRP). The current effort to finish IEEE 802.15.3c and ECMA standards for 60 GHz systems demonstrate the interest on the side of the industry for this frequency range [1.8], [1.9].

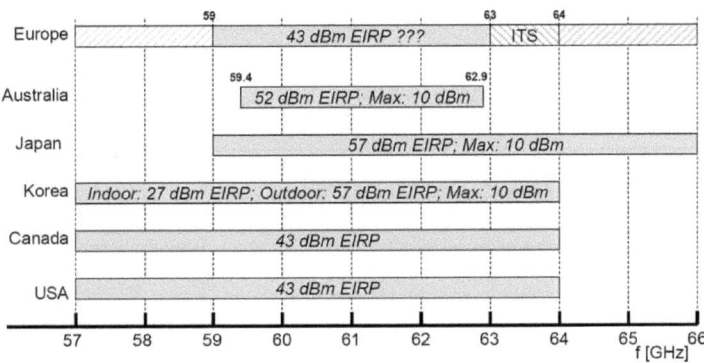

Figure 1. 2 Spectrum allocation in the unlicensed 57-66 GHz band.

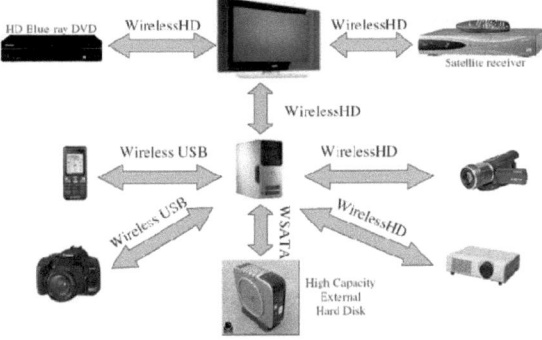

Figure 1. 3 Some possible applications for 60 GHz radio.

The 60 GHz radio is envisaged for usage in a broad spectrum of wireless personal area networks (WPANs) and wireless local area networks (WLANs). Some of the possible applications are shown in Figure 1.3. Others include entertainment systems on-board of aircrafts, trains and cars. A commercially very attractive application is wireless connection for high-definition TV (HDTV).

1.2. IHP's SiGe:C BiCMOS Technology

An IHP's SiGe technology was used to develop all the circuits of the analog front-end (AFE) and for the final integration of the transmitter and receiver in the WIGWAM and EASY-A projects. It is a SiGe:C BiCMOS 0.25 µm technology and it was presented in [1.10]. When the work on the AFE started it was the fastest IHP technology and presented the state-of-the-art in the SiGe technologies. The progress achieved in increasing the speed of SiGe heterojunction bipolar transistors (HBTs) made SiGe technologies competitive with III/V technologies in the 60 GHz range and above [1.11].

Component	Parameter	Value
NPN HBT	Peak f_T	180 GHz
NPN HBT	Peak f_{max}	220 GHz
NPN HBT	β	200
NPN HBT	BV_{CE0}	1.9 V
NPN HBT	BV_{CB0}	4.5 V
NPN HBT	V_A	40 V
MIM Capacitor	C	1 fF/µm²
Resistors R_{sil}, R_{pnd}, R_{high}	R_{sheet}	7, 210, 1600 Ω/□

Table 1.I Overview of IHP's SiGe:C BiCMOS technology.

The main parameters of the used HBTs and other components are presented in Table 1.I. The peak transit frequency (f_T) of 180 GHz and the peak maximum frequency of oscillation (f_{max}) of 220 GHz are sufficient for the design of 60 GHz AFE circuits, but with f_T being just three times the working frequency, the performance of the chips in terms of achievable gain and output power is limited.

The transistors are available in size with 1 to 8 fingers. Size one transistor has drawn emitter dimensions 0.18×0.84 µm². The used transistor model is VBIC (Vertical Bipolar Inter-Company model). Silicon nitride (SiN) MIM capacitors are used with capacitance of 1 fF/µm². Three types of polysilicon resistors are available with sheet resistance of 7, 210 and 1600 Ω/□. There are five aluminium metal layers. The top metal layer (TM2) is optional and has thickness of 3 µm. It is used for high-Q passive structures. The chips were produced with this metal layer. A varactor with MOS structure is also available. It is used in the design of voltage control oscillators (VCOs).

1.3. Thesis Objective and Organization

The work presented in this dissertation is largely related to my work in the WIGWAM and EASY-A projects. The goal of these projects was the development of 60 GHz wireless communication demonstrators with fully integrated transmitter (TX) and receiver (RX) chips fabricated in IHP's SiGe technology. My tasks in the beginning were development of a power amplifier (PA), and image-rejection filter and the optimization of a phase-locked loop (PLL) (the PLL was earlier developed). Later, my work shifted to TX integration and AFE tests.

The state-of-the-art at the beginning of the work was such that there were no reported integrated TXs or RXs working at 60 GHz. The first reported demonstrator from the WIGWAM project was worlds second [1.12], [1.13]. However, there was more work done related to specific 60 GHz AFE components [1.4].

The objectives of this thesis stem mostly from the project objectives, but partly include broader theoretical analysis, which is not related to the project goals. The goal of the thesis is the analysis of the challenges related to the design of mm-wave transceivers, and design of a 60 GHz TX which improves the state-of-the-art. The objective for the PA is the analysis of the challenges related to the design of integrated SiGe mm-wave power amplifiers, and development of a PA, which improves the state-of-the-art. PA design is very important because it significantly affects the performance of the whole TX. The objective for the image-rejection filter is the analysis of achievable results for integrated mm-wave filters, and development of a filter with sufficient image-rejection. The objective for the PLL is optimization of the PLL noise performance with respect to the performance of the whole TX and RX. The dissertation additionally deals with the optimization of the PLL parameters, which is not related to the project goals. It presents comparison of different PLL topologies and introduces a new approach for their calculation.

Since the AFE design was a team work it is sometimes hard to draw a line between my own work and the work of my colleagues. Furthermore, neither is my work restricted to the TX (the PLL is also used in the RX), nor is the TX solely my work (the upconversion mixer and the first version of the PLL were done by my colleagues). The results which are not my own, but are important for the topic will be only briefly presented.

In addition to presenting results of my work, the dissertation contains short theoretical overview for the topics of my work (PLL, image-rejection filter and PA). The presented theory is selective in sense that it is relevant to the specific design. Other topics such as mixer design or transmitter topology are not accompanied with relevant theory section.

The next chapter deals with the AFE and specifically with the transmitter architecture. Different versions of the AFE are presented. The chapter also presents the required parameters for the TX components: the PLL, image-rejection filter and PA. Other TX components (the upconversion mixer and preamplifier) are also briefly presented.

The third chapter presents the relevant PLL theory, PLL optimisation, the new approach in calculating the PLL parameters, the PLL design and the measurement results.

The fourth chapter deals with the image-rejection filter. The theory and the design of different filter types are presented with their measurement results.

The fifths chapter introduces the PA design theory, discusses different mm-wave PA design issues and presents the measured results.

The sixth chapter presents measurement results for different versions of the transmitter.

The last chapter summarizes the presented work and its results.

Chapter 2
Transmitter Architecture

2.1. Introduction

This chapter presents the main performance requirements for the 60 GHz wireless communication system that was developed. The envisaged application scenarios are briefly presented. The effect of the performance requirements on the TX and RX architecture is presented. Different versions of the TX that were developed are also presented and the TX topology choice is discussed. The required performance of the TX is analysed and the performance parameters of the TX components are calculated. The upconversion mixer and the preamplifier are presented.

2.2. Application Scenarios and TX Requirements

The 60 GHz AFE was envisaged for different kinds of application scenarios. They entail WLAN, video streaming, file download and others. One example is an entertainment system on-board a plane. Another example is a kiosk for sales of large files such as movies or games, which would be downloaded to a portable storage device.

System level calculations and simulations were done to estimate the needed performance parameters for the AFE, which would enable the desired performance for the system.

The chosen modulation is orthogonal frequency division multiplex (OFDM). It has important advantages of high spectral efficiency and robustness to multipath. Disadvantages include relatively complex AFE and baseband (BB) (compared with single carrier modulation), and the main disadvantage is high peak-to-average power ration (PAPR).

System level simulations have given two main requirements for the transmitter. One is supplying 10 dBm output signal at 60 GHz with low level of nonlinearities. The other is low level of PLL phase noise. The image–rejection should be at least 30 dB.

2.2.1. Link Budget Calculation

The maximum distance of communication for a given system is limited by the minimum signal-to-noise ratio (SNR_{min}) of the received signal, which is sufficient for its detection. The value of SNR_{min} depends on several factors, but the main are the used modulation scheme and the maximum acceptable error rate (usually bit/symbol error rate or FER). For a LOS system, the maximum distance can be calculated from the link budget:

$$P_{RXin}[dBm] = P_{TXout} - L_{TX} + G_{ATX} - L_{FS}(d) + G_{ARX} - L_{RX} \quad (2.1)$$

where P_{RXin} is received signal power (in dBm); P_{TXout} is transmitter output power (in dBm); L_{TX} are transmitter losses (in dB); G_{ATX} is transmitter antenna gain (in dBi); $L_{FA}(d)$ is free-space path loss and from Equ. (1.1) at 60 GHz $L_{FA}(d) = 68dB + 20\log(d[m])$; G_{ARX} is receiver antenna gain (in dBi) and L_{RX} are receiver losses (in dB).

Thermal noise level (N_0) at the input of the receiver at temperature T (in K) is given as: $N_0 = kT$, where k is Boltzmann's constant (k = 1.38×10^{-23} J/K). At room temperature T = 290 K, the thermal noise level is –174 dBm/Hz. The power of the noise in the bandwidth of interest B is given as:

$$P_n[dBm] = 10\log(kTB) = -174dBm + 10\log(B[Hz]) \quad (2.2)$$

If F is the RX noise figure, the SNR of the received signal in the presence of only thermal noise is given as:

$$SNR[dB] = P_{RXin} - P_n - F \quad (2.3)$$

In reality the transmitted signal is not ideal and has some noise ($P_{N,TX}$ – level at the input of RX), giving the total SNR of the received signal:

$$SNR[dB] = 10\log(P_{RXin}/(P_{N,TX} + kTBF)) \quad (2.4)$$

Finally, for maximum transmission distance d_{max}, we get minimum SNR:

$$SNR_{min}[dB] = P_{TXout} - L_{TX} + G_{ATX} - 68dB - 20\log(d_{max}) + G_{ARX} - L_{RX} - 10\log(P_{N,TX} + kTBF) \quad (2.5)$$

2.3. TX Topology

The AFE was produced in two versions. In both cases it features heterodyne topology, but with different intermediate frequency (IF). The heterodyne topology is widely used in wireless transceivers. It is more complex then the homodyne topology, but doesn't share the problems related to DC offset, LO leakage and even-order distortions, which are common in homodyne transceivers.

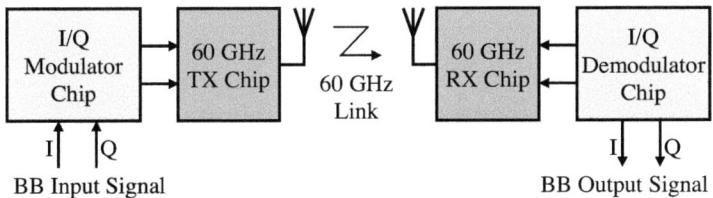

Figure 2. 1 AFE block diagram.

The block diagram of the first version of the AFE is shown in Figure 2. 1. The baseband provides I and Q signals for the first chip – an I/Q modulator. The differential output is fed into a 60 GHz transmitter. On the receiver side, the signal is first downconverted in a 60 GHz receiver chip, and then in an I/Q demodulator. The IF frequency in the first version is 5.25 GHz (for simplicity, it will be later in text referred to as 5 GHz). It was chosen for compatibility with IEEE 802.11a standard, so that the 60 GHz chips can be used with commercial UWB systems. The PLL frequency used for the second upconversion (in the 60 GHz TX and RX chips) is 56 GHz. Signal bandwidth is 360 MHz (I and Q have 180 MHz bandwidth each).

In the second version of the AFE, the IF chips and 60 GHz chips are integrated giving one TX and one RX chip. The AFE features a modified heterodyne topology – called sliding-IF. It allows using only one PLL with a VCO working at 48 GHz. The IF frequency is generated dividing the VCO signal by four (12 GHz). When a PLL reference is changed both the PLL and IF frequencies change (slide) – giving the name sliding IF. Signal bandwidth is 1.7 GHz (I and Q have 850 MHz bandwidth each). In addition to having just one PLL, this topology is advantageous because high IF makes the design of IF circuitry easier for such broadband input signals.

The upconversion to the IF frequency of 5 GHz (12 GHz for second version) uses quadrature modulation for image suppression. Details of the 5 GHz IF chips are presented in [2.2]. The second upconversion is done in a single mixer with a 56 GHz (48 GHz) signal. The result is double-sideband spectrum with signal at 61 GHz (60 GHz) and out-of-band image at 51 GHz (36 GHz), which has to be attenuated.

The 60 GHz TX is realised in three versions. The first two are for the first version of the AFE with 5 GHz IF. The third TX version is for the second AFE version with 12 GHz IF.

2.3.1. Version I TX Topology

The block diagram of the version I TX is show in Figure 2. 2. The TX consists of an upconversion mixer, phase-locked loop, a preamplifier, an image-rejection filter and a power amplifier. The mixer, PLL, preamplifier and PA are differential. The image-rejection filter is single-ended. Since the first version of the antenna was single-ended, the TX was not designed as fully differential. It was simplified by terminating (on-chip) one preamplifier and one PA output with 50 Ω, and using only one image-rejection filter. Since the filter is relatively large, the chip is smaller.

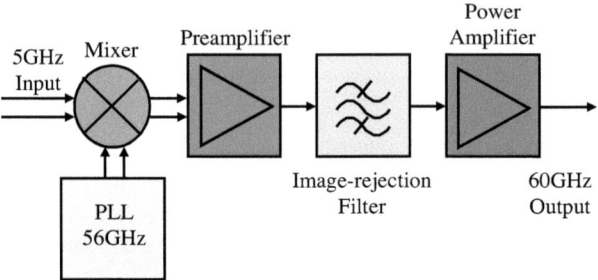

Figure 2. 2 Transmitter version I block diagram.

2.3.2. Version II TX Topology

The block diagram of the version II TX is show in Figure 2. 3. This version is very similar to the version I, except that it is fully differential. Two image-rejection filters are implemented for the differential signal, and the output signal is fed into a differential on-board Vivaldi antenna.

2.3.3. Version III TX Topology

The block diagram of the version III TX is show in Figure 2. 4. This is a fully integrated TX, which includes I and Q mixers for IF quadrature upconversion, an IF amplifier, an upconverstion mixer, two image-rejection filters and a PA.

IF circuitry was redesigned for 12 GHz IF frequency and for broadband I and Q input signals (850 MHz bandwidth each). The gain in the IF stage and sideband suppression of I and Q mixers can be controlled with an integrated SPI.

In contrast to the first two TX versions, the preamplifier was here omitted. High enough IF signal level, mixer output P1dB and low filter insertion loss made the omission possible.

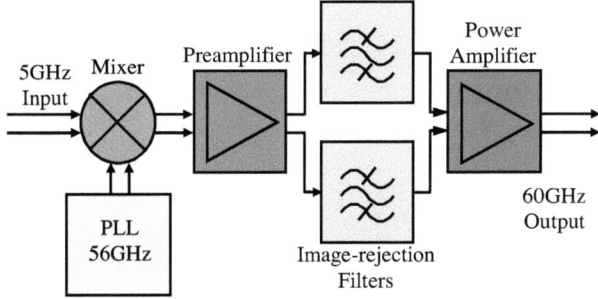

Figure 2.3 Fully differential transmitter version II block diagram.

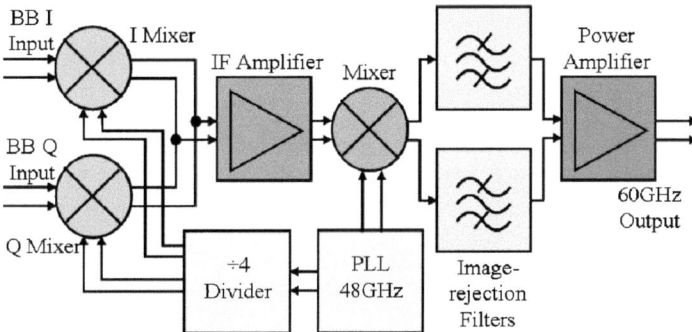

Figure 2.4 Fully integrated differential transmitter version III block diagram.

2.3.4. Upconversion Mixer

The IF signal from the I/Q modulator is upconverted in the mixer with the 56 GHz oscillator signal from the integrated PLL. The result is a double-sideband spectrum with a signal at 61 GHz and an out-of-band image at 51 GHz.

The upconversion mixer features the Gilbert cell topology. The mixer is optimized for high linearity and output power. To achieve a high voltage swing at the output, a large resistor value is required for the load resistor R_C. However, the voltage drop across the resistor increases with increasing load resistance. This will deteriorate the transistor performance. A 200 Ω load resistor represents a good trade-off. When the 5 GHz signal is upconverted to 61 GHz, there is a conversion loss. The maximum output P1dB while using the smallest transistors is −15 dBm, −12 dBm differential. The simplified schematic of the mixer is shown in Figure 2.5. Transistors Q1 to Q6 together with two load resistors R_C form the Gilbert cell core.

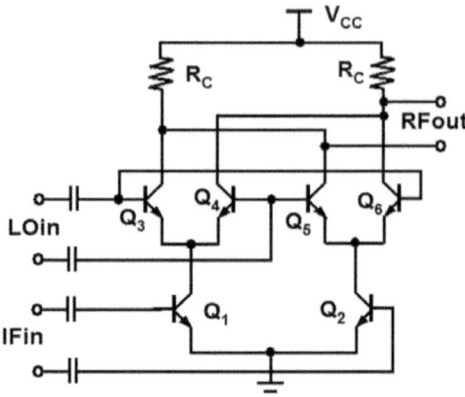

Figure 2. 5 Simplified schematic of the upconversion mixer.

The mixer draws 10 mA from a 3.5 V supply. For optimal performance, the mixer requires a 4 dBm differential input signal at 56 GHz from the PLL.

This mixer was designed by a colleague and it was used for TX, version I and II, with 5 GHz IF. The mixer for TX version III with 12 GHz IF was my design. It was essentially a redesign of the first version and has the same schematic as in Figure 2. 5. The redesign had two goals:

1. To optimize the frequency characteristic for 12 GHz input.
2. To increase the output P1dB to −7 dBm, −4 dBm differential. This was done to omit the preamplifier, because with stronger signal at the output of the mixer and less insertion loss in the image–rejection filter the signal at the input of the PA was sufficient to drive it without degrading output P1dB.

The output power of the mixer was increased by using the new faster HBT transistors at the cost of higher power consumption. The load resistor R_C is 140 Ω. The new mixer draws 25 mA from a 3.3 V supply. Higher power consumption of 47 mW pays off because the omitted preamplifier consumes 156 mW.

2.3.5. Preamplifier

The preamplifier design is based on the 60 GHz low-noise amplifier (LNA) design presented in [2.1]. The LNA was designed by a colleague, and the preamplifier was my design. The purpose of the preamplifier is to provide additional gain for the 60 GHz signal. This is needed because the mixer P1dB is low, and the image–rejection filter has relatively high insertion loss.

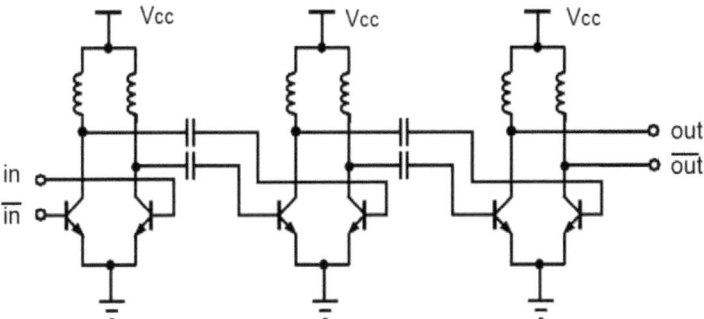

Figure 2.6 Simplified schematic of the preamplifier.

The preamplifier features a three-stage common-emitter differential topology. Figure 2.6 shows the simplified schematic. Inductors for matching are realized as metal lines. The loss of microstrip transmission lines (TLs) is high due to the relatively thin silicon dioxide layer between the top metal and ground metal. When this circuit was designed only four metal layers were available, so metal4 was used for TLs. Later metal5 was made available (called top-metal1 TM1). In order to obtain a symmetrical design, a single-ended preamplifier was laid out first, and then it was copied upside down. The emitters of the differential pairs are connected directly to the chip ground. By doing so, the preamplifier is pseudo differential and single-ended measurable, but it has no common-mode rejection.

It has a measured gain of 18 dB at 61 GHz and output P1dB of 3 dBm. The preamplifier consumes 60 mA from a 2.6 V supply. The frequency characteristic is broadband resulting in minor image-rejection capability of 2–to–3 dB.

The preamplifier was used in TX version I and II, and it was omitted in version III.

2.4. Summary

This chapter presented the main application scenarios and AFE versions. Three TX versions were presented. Two TX components, the upconversion mixer and preamplifier were described. Based on this data, and on system level calculations and simulations the following requirements are derived:

PLL. In order to allow detection in the RX with low BER, PLL phase noise must be optimized. The RMS phase error of the PLL signal has to be minimized. The simulations have shown that the RMS phase error after common phase error (CPE) correction has to be below 3 degrees to achieve the desired performance.

The system is not sensitive to the PLL spurs, but they should be below –30 dBc, and they can be neglected if they are below –40 dBc.

Image–rejection filter. The image-rejection should be at least 30 dB. For the second version of the AFE with IF 12 GHz, this is not difficult to achieve. For the first version with 5 GHz IF, this is difficult, so the requirement was split to 20 dB image–rejection for the filter and 10 dB for the PA.

PA. In order to amplifier an OFDM modulated signal without significant amplitude distortion, the signal at the output has to be several decibels below the 1 dB compression point (P1dB) of the amplifier. The simulations showed that for 16-QAM modulation the back-off from P1dB should be 5 dB for frame error rate (FER) below 1 %. For QPSK the back-off can be just 1 dB for FER below 1 %. Introducing back-off is not desired because it brings loss in the link-budget and low power efficiency of the transmitter power amplifier.

The PA has to to be optimized for high linear output power. The required P1dB is at least 10 dBm. More is highly desirable because it directly improves the performance of the whole communications system. Due to the low mixer output P1dB and filter insertion loss, the gain has to be at least 25 dB. Several dBs more of gain is an advantage – they serve as a reserve against gain drop due to process variation or high temperature. The PA should be selective at the image frequency and provide at least 10 dB of image rejection.

Chapter 3
Phase–locked Loop

3.1. Introduction

This chapter presents the phase-locked loops (PLLs), which are used in the AFE version I and II. The PLL for the AFE version I (56 GHz) is an optimized design from a PLL which was designed by a colleague. The first section presents basic theory of the used PLL. Issues of the PLL functionality and stability are addressed together with the important issue of noise performance of a PLL. The calculation of the PLL bandwidth and phase margin for the minimum RMS phase error is presented.

The following section presents existing recipes for PLL parameter calculation. A new recipe for the forth-order PLL parameter calculation is presented. It may be used for achieving better spur suppression or for calculation of PLL parameters, which are easier for implementation, such as smaller LPF capacitors or lower charge-pump current.

Different PLL topologies are presented in the fifth section. These topologies are compared for their phase noise performance and spur suppression. We also analyze which topology is optimal for certain design requirement.

PLL measurement results are presented, as well as measured noise performance of the whole system, which is mainly influenced by the 56/48 GHz PLL.

3.2. PLL Theory

Phase–locked loops have broad and diverse usage. Most important applications are: generation of signals with stable frequency, recovery of signals from a noisy communication channel and clock distribution in digital logic designs. PLLs are used in transceivers to deliver a clean (low phase noise), stable signal with well defined frequency for up- or downconversion. There are two main analog PLL types. A PLL type I uses a phase detector for phase comparison, and a PLL type II uses a phase-frequency detector and a charge pump. PLL type II is used more broadly today, because it has better locking capabilities. Since both the 56 GHz PLL used in AFE version I and the 48 GHz PLL used in AFE version II are type II, the theoretical analysis will focus on this type.

3.2.1. Type I PLL

Basic PLL is presented in Figure 3. 1. This is a type I PLL. It consists of a phase detector (PD), a lowpass filter (LPF) and a voltage-control oscillator (VCO). It may additionally have a divider, shown in Figure 3. 1. as a block N, where n represents the division ratio. Before explaining the working principle of a PLL, let us first examine the function of each component.

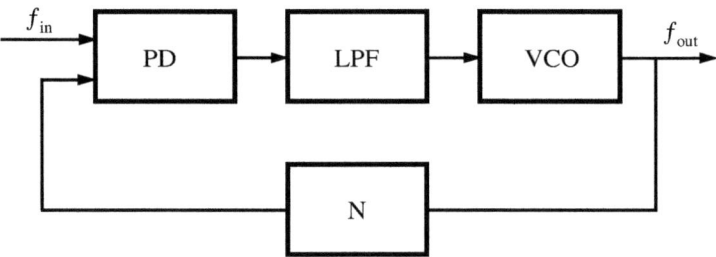

Figure 3. 1 Basic type I phase–locked loop.

3.2.1.1. Type I PLL Components

Phase detector is a component with two inputs and one output. The signal at the output is a dc voltage directly proportional to the phase difference ($\Delta\varphi$) of the two input periodical signals, i.e.:

$$v_{out} = K_{PD} \Delta\varphi \qquad (3.1)$$

where K_{PD} represents the gain of the phase detector (in V/rad). The PD input-output transfer function in s-domain (after Laplace transformation) is a constant:

$$\frac{V_{out}}{\Delta\Phi}(s) = K_{PD} \qquad (3.2)$$

A phase detector is usually a multiplier, which is realized as a Gilbert cell. The transfer function of such phase detector is the sinusoidal function ($v_{out} = K_{PD} \cos(\Delta\varphi)$), which is a good approximation of the ideal linear function for $\Delta\varphi \approx \pi/2$.

Lowpass Filter. A simple lowpass filter, which is used for the type I PLL analysis here, is shown in Figure 3. 2. The voltage transfer function of this filter in *s*-domain is given by:

$$F(s) = \frac{1}{1 + \frac{s}{\omega_p}} \tag{3.3}$$

where $\omega_p = 1/(RC)$. We should note here, that for a type II PLL, the signal at the input of the filter is current, and the output is voltage resulting in transimpedance transfer function ($V_{out}/I_{in}(s)$).

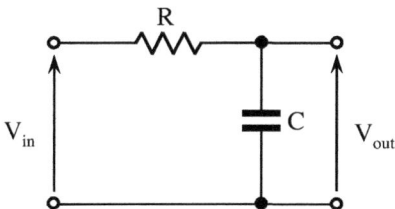

Figure 3. 2 Simple lowpass filter for the type I PLL.

Voltage-controlled oscillator. An oscillator with the output frequency, which can be varied by a voltage is called a voltage-controlled oscillator. In LC oscillators this variation is usually done by introducing a variable capacitance (usually a reverse-biased diode – varicap) in the oscillator tank. The output signal frequency of an ideal VCO is a linear function of a control voltage (V_{ctrl}):

$$\omega_{out} = \omega_0 + K_{VCO} V_{ctrl} \tag{3.4}$$

where ω_0 is the "free-running" frequency (or frequency for $V_{ctrl} = 0V$), and K_{VCO} is the "gain" of the VCO (in rad/V). The angular frequency ω_0 may not have any physical meaning. It is present in the equation (3.4) because for the range of V_{ctrl} the output frequency is not zero.

The phase of the output signal is the integral of frequency with respect to time, so it can be expressed by:

$$\varphi_{out}(t) = \omega_0 t + K_{VCO} \int_{-\infty}^{t} V_{ctrl} dt = \omega_0 t + K_{VCO} \int_{0}^{t} V_{ctrl} dt + \varphi_0 \tag{3.5}$$

where φ_0 is the phase for $t = 0$. The transfer function of a VCO for the input signal $V_{ctrl}(t)$ and output signal $\varphi_{out}(t)$ in *s*-domain is given by:

$$\frac{\Phi_{out}}{V_{ctrl}}(s) = \frac{K_{VCO}}{s} \tag{3.6}$$

Divider. A divider, with a division ratio *n*, for a periodic input signal, with frequency f_{in}, produces a periodic signal at the output with frequency $f_{out} = f_{in}/n$. The integration of this equation gives the relation for the input and output signal phase:

$$\varphi_{out}(t) = \frac{\varphi_{in}(t)}{n} + \varphi_0 \qquad (3.7)$$

where φ_0 is the phase for $t = 0$. φ_0 represents the phase delay of the divider. The phase transfer function of a divider in s-domain is given by:

$$\frac{\Phi_{out}}{\Phi_{in}}(s) = \frac{1}{N} \qquad (3.8)$$

3.2.1.2. Type I PLL Operating Principle

A phase-locked loop is a feedback system. Unlike usual feedback systems which operate on the voltage, the PLL operates on the excess phase of periodic signals. The phase detector generates a signal whose dc voltage is proportional to the phase difference of the PLL input signal and VCO output signal. It means that it converts the excess phase signal into a dc voltage. The LPF filters the PD output signal, and provides the dc voltage, which controls the oscillating frequency of the VCO directly, and the signal phase indirectly. We may see that the loop signal is converted from the phase to dc voltage and back to the phase.

The PLL may be in a dynamic or static state. Figure 3.3 shows PLL signals when the PLL is in a static state. The phase difference between the PLL input signal and the PLL (VCO) output signal ($\Delta\varphi$) is then constant. When this phase difference is constant, the PLL is said to be *locked* (hence the name: phase-locked loop). In that case, the phase detector gives a signal with constant dc value. The PD signal is filtered by the LPF. The output LPF dc voltage (i.e. PD dc voltage) determines the frequency of the VCO signal. The VCO signal frequency is constant resulting in constant phase difference, $\Delta\varphi$.

Constant phase difference between the input and output signal means that the VCO oscillates at the frequency of the input signal. If the PLL has a divider with division ratio n, the VCO will oscillate at the frequency n times higher than that of the input signal, i.e. $f_{out} = nf_{in}$.

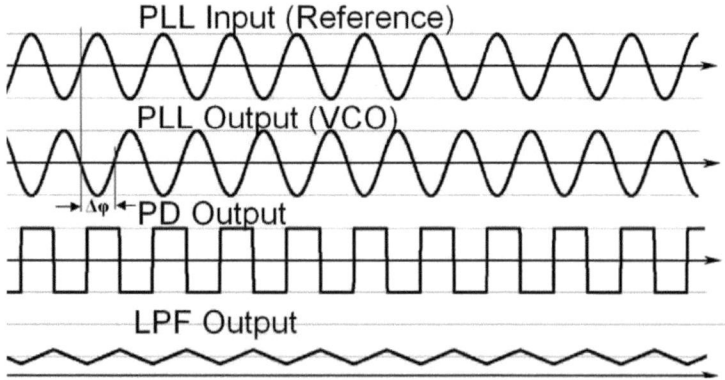

Figure 3.3 Type I PLL signals when the PLL is locked.

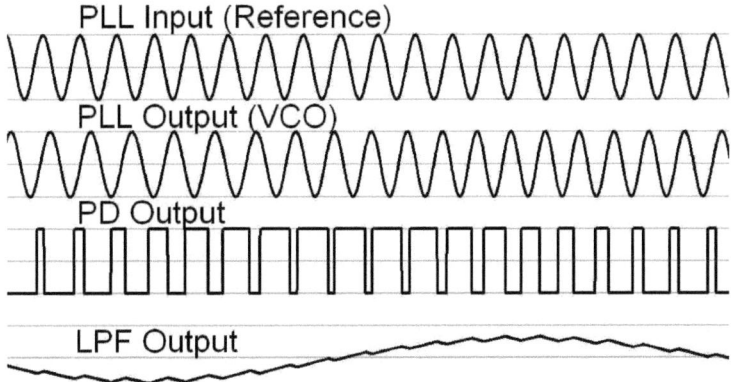

Figure 3. 4 Type I PLL signals while the PLL is locking.

The phase difference of the locked state $\Delta\varphi$ depends on the input frequency and PD architecture. If $f_{out} = nf_{in}$ for $V_{ctrl} = 0$, then the phase difference will depend only on the PD architecture. The phase detector used to generate signals in Figure 3. 3 has digital output with values $\pm V_0$. The phase difference in Figure 3. 3 is 180° because the PD in that case gives signal with zero dc component, which doesn't change the LPF output and the VCO frequency. If a PD signal has values 0 and V_0, the locked state phase difference would be zero, so no V_0 peaks would be created and the dc voltage would be zero. A PD realized as a multiplier with Gilbert cell has an analogue output signal, which equals zero for input phase difference of 90° (as explained in the PD section).Let us now examine what happens when the input frequency is changed by a small percentage. The input frequency and the VCO frequency are now different and the phase difference at the PD input grows with time. The PD starts generating signal with a different dc value and it changes the LPF output dc signal. This in turn changes the VCO frequency. In other words, the PLL (VCO) frequency tracks the input frequency changes. This process is hence called tracking. Figure 3. 4 shows PLL signals during this process.

The PLL is back in the static state only when both the input and output frequency are equal and the phase difference $\Delta\varphi$ has the proper value (which is in our case 180°) [4.1]. During the process of locking, the input and output frequency may be equal at certain point in time, but if the phase difference is not right, the PLL will again change the VCO frequency, for a while, in order to compensate for the phase difference.

When the phase difference $\Delta\varphi$ varies with time, the PLL is not locked. In the unlocked state we differentiate two situations. One happens when the PLL is locked and the input frequency changes a little. The PLL adjusts to the input change by following or tracking the input frequency. The second situation happens when the input frequency changes a lot (i.e. fast increase in phase difference), or when the PLL has just been switched on. The PLL is then in the state of acquisition. It is clear that the PLL cannot lock to the input signal, if the input frequency is out of the VCO range. The VCO range is determined by the maximum dc voltage span that the PD can generate at the output of the LPF.

The question if and how the PLL locks is essential for PLL usage. The process of locking is a nonlinear phenomenon, which is, as such, difficult to analyze. Type I PLL acquisition is not of our interest and will not be discussed here. More detailed discussion of this phenomenon is given in [3.1-3]. The most important result of the analysis is that type I PLLs have a smaller acquisition range than the tracking range [3.1].

However, transient response of a PLL can be analyzed using a linear approximation and this analysis is important to understand PLL functioning.

3.2.2. Type II PLL

The main disadvantage of the type I PLL is that the acquisition range is smaller than the tracking range. Type II PLL (shown in Figure 3. 5) has a phase/frequency detector (PFD) and a charge pump (CP) instead of the phase detector. PFD detects both the frequency and phase and enables the PLL to lock to any input frequency within the VCO range (provided that the PLL can supply the required VCO V_{ctrl} voltage).

3.2.2.1. Type II PLL Components

Phase/Frequency Detector. The PFD compares the two periodic input signals (A and B) and provides two outputs (Q_A and Q_B). If the signal A angular frequency, ω_A, is larger than the signal B angular frequency, ω_B, output Q_A supplies signal with dc component, which is proportional to the phase difference of the input signals. Q_B output is equal to zero. If ω_A is smaller than ω_B, Q_B output supplies signal with dc component, which is proportional to the input phase difference and Q_A output is equal to zero (see Figure 3. 6).

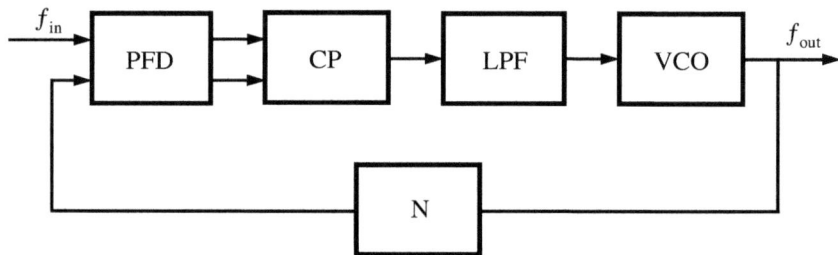

Figure 3. 5 Basic type II phase–locked loop.

An ideal PFD is a state machine with state diagram shown in Figure 3. 7. Rising A and B edges change the PFD state and output signals. When $\omega_A > \omega_B$, the PFD toggles between state 0 and I. Signal A rising edge shifts the PFD from state 0 to state I, and signal B rising edge shifts it back to state 0. If signal A rising edge comes while the PFD is already in state I (which happens when ω_A differs a lot from ω_B) – the PFD will stay in the same state. This is an important fetcher because it enables the PFD to "detect" not just the phase, but also the frequency and keep generating signals to increase the VCO frequency as long as $\omega_A > \omega_B$. Q_A output is also called "UP" signal (because it is active while the PLL output frequency, ω_B, is lower than the input frequency, ω_A). Q_B output is called "DOWN" signal.

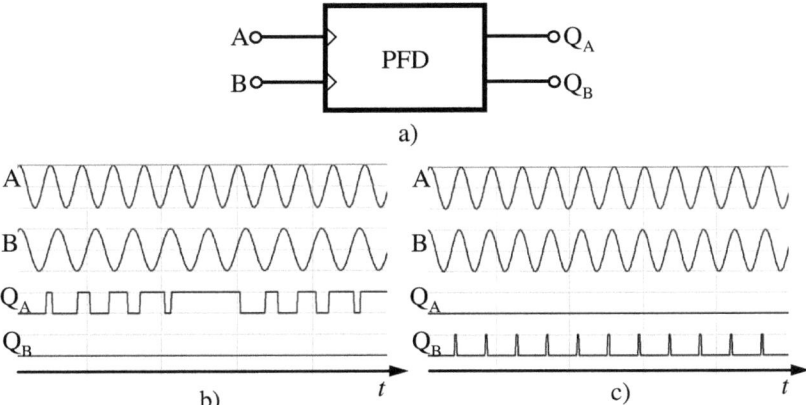

Figure 3.6 PFD symbol a). PFD response for $\omega_A > \omega_B$ b), and $\omega_A = \omega_B$ with ω_A lagging c).

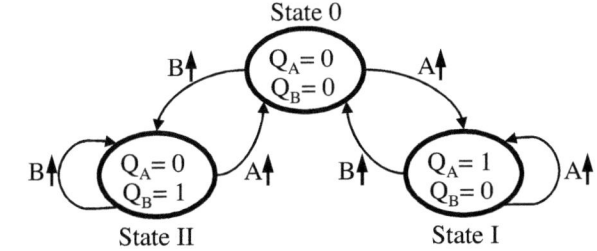

Figure 3.7 PFD state diagram.

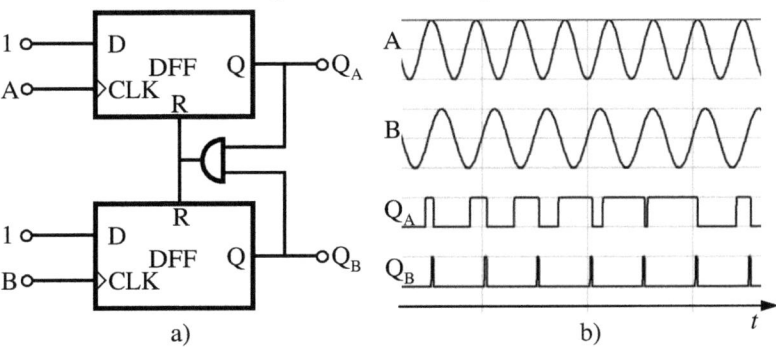

Figure 3.8 PFD implementation.

Figure 3.8 shows a usual PFD realization with two D flip-flops and an AND circuit. We may note one difference between the real and ideal PFD. The real PFD generates short signals on the "inactive" output (see Figure 3. 8b). The length of these impulses is the time delay thru the AND circuit and the time to reset the DFFs.

Charge Pump. The model of an ideal charge pump is shown in Figure 3. 9. The UP (Q_A) and DOWN (Q_B) signals from the PFD are used to control the switches S_1 and S_2. When the UP signal is active, the switch S_1 is closed and current I_1 is flowing to the output, charging the loop filter capacitor, i.e. increasing the VCO control voltage. When the DOWN signal is active, I_2 flows out, decreasing the VCO control voltage. When the UP and DOWN signals are not active, there is no charging or discharging of the loop filter.

When the PLL is locked, phase difference of the PFD input signals is constant. For the type I PLL, the value of the phase difference depends on the input frequency and PD type, as explained in section 3.2.1.2. In the case of a type II PLL (charge-pump PLL), phase difference has to be zero. Any other value would change or discharge the loop filter and change the VCO control voltage.

As mentioned in the previous section, a real PFD generates UP and DOWN impulses with certain minimal width. These impulses have to be long enough to switch the switches. If they are too short, the charge pump will not function for small phase differences, and the PLL will not be able to lock or to stay locked. This problem is known as the "dead zone".

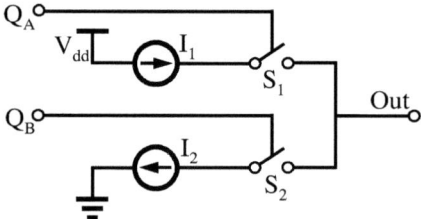

Figure 3. 9 Ideal charge pump model.

3.2.2.2. Type II PLL Linear Model

A type II PLL, or charge-pump PLL (CPPLL), is a discrete-time system, because the CP works on a switching principal and there is no discharge path between phase comparison instants. However, if the loop bandwidth is much smaller than the input frequency, we can assume that the PLL state changes by a small amount during each circle of the input [3.7]. By averaging the value of the discrete–time parameters, the PLL can be analyzed as a continuous–time system [3.7].

A linear model of the locked PLL is shown in Figure 3. 10. The PFD can be represented with a block for phase subtraction. The UP signal represents positive phase difference, and the DOWN signal negative phase difference. The charge pump corresponds to a block with constant gain $I_{CP}/2\pi$, where I_{CP} is the charge pump current ($I_{CP} = I_1 = I_2$).

The LPF transfer function of interest is transimpedance, i.e. current-to-voltage transfer function $F(s)$. The VCO transfer function is given as K_{VCO}/s (see Equ. (3.6)).

Chapter 3 Phase–locked Loop

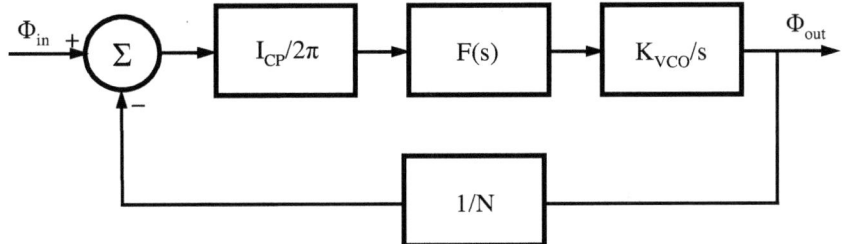

Figure 3. 10 Type II PLL linear model.

The open loop transfer function, $H_0(s)$ is given by [3.4]:

$$H_0(s) = \frac{I_{CP}}{2\pi} F(s) \frac{K_{VCO}}{s} \qquad (3.9)$$

and the closed-loop transfer function $H(s)$ is given by:

$$H(s) = \frac{H_0(s)}{1 + \frac{H_0(s)}{N}} \qquad (3.10)$$

The simplest LPF is a shunt capacitor (see Figure 3. 11a) with the transfer function $F(s) = 1/sC$. A PLL with such LPF would have two poles in the origin (one from the LPF and the other from the VCO) in the open-loop transfer function:

$$H_0(s) = \frac{I_{CP} K_{VCO}}{2\pi C_p s^2} \qquad (3.11)$$

resulting in 0° phase margin and loop instability. To avoid this, a zero must be added in the open-loop transfer function. This can be done by introducing a resistor R in the LPF (see Figure 3. 11b).

The closed-loop transfer function is then given by:

$$H(s) = \frac{\frac{I_{CP}}{2\pi C_p}(1 + sRC_p) K_{VCO}}{s^2 + \frac{I_{CP}}{2\pi N} K_{VCO} Rs + \frac{I_{CP}}{2\pi C_p N} K_{VCO}} \qquad (3.12)$$

Equation (3.12) can be converted so that the denominator has form $s^2 + 2\zeta\omega_n s + \omega_n^2$, which is a form used in the control theory. The equation (3.12) becomes:

$$H(s) = \frac{\omega_n^2 N}{s^2 + 2\zeta\omega_n s + \omega_n^2} \qquad (3.13)$$

where ω_n is the natural frequency of the system and ζ is the damping factor, given by:

23

$$\omega_n = \sqrt{\frac{I_{CP}}{2\pi C_p N} K_{VCO}}$$

$$\zeta = \frac{R}{2}\sqrt{\frac{I_{CP} C_p}{2\pi N} K_{VCO}} \quad (3.14)$$

Some applications require large loop bandwidth. The bandwidth is usually proportional to ω_n, and it is increased with higher I_{CP} and K_{VCO}. If the loop bandwidth is comparable with the input frequency, the continuous–time approximation is no longer valid, and the discrete–time analysis is required. A stability limit for the PLL natural frequency ω_n for an input angular frequency ω_{in} is given by [3.7]:

$$\omega_n^2 < \frac{\omega_{in}^2}{\pi(RC_p \omega_{in} + \pi)} \quad (3.15)$$

This equation implies that R, although it doesn't affect ω_n directly, cannot be increased indefinitely. In typical designs, the loop bandwidth is about one–tenth of the input frequency to insure stability.

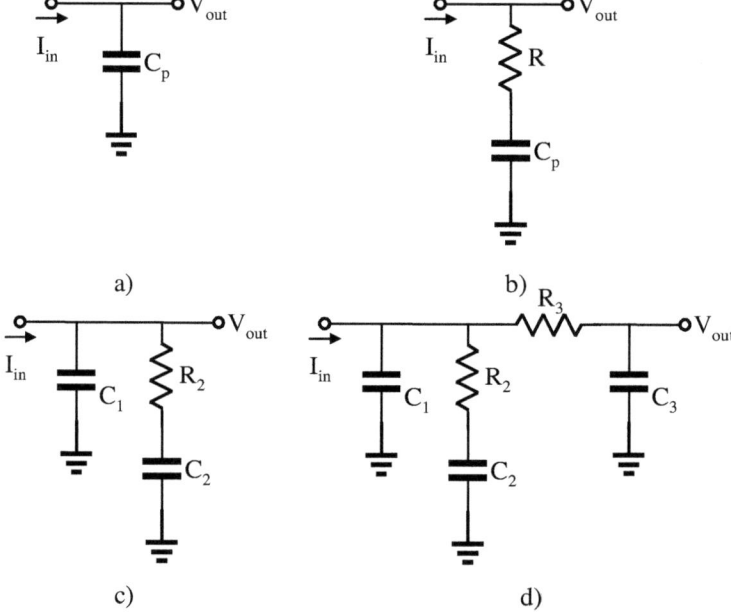

Figure 3.11 a) The simplest type II PLL LP filter. b) Type II PLL filter with zero in transfer function. c) Type II PLL filter with the ripple suppression capacitor C_1 (third order PLL). d) Type II PLL filter with the ripple suppression filter stage $R_3 C_3$ (fourth order PLL).

In real PLLs, the CP switches S_1 and S_2, are turned on at every phase comparison instance, and the mismatch between I_1 and I_2 flow into the LPF. This current introduces ripple on the resistor in the LPF (see Figure 3. 11b) in the VCO control voltage. The ripple modulates the VCO output frequency, which is especially undesirable in frequency synthesizers.

The ripple can be suppressed by adding an additional capacitor in the LPF, as shown in Figure 3. 11c. This capacitor introduces additional pole in the PLL transfer function and such a PLL is called a third order PLL.

3.2.2.3. Type II PLL Stability Analysis

A PLL is a system with a feedback, and as such can oscillate. From linear systems theory, a linear system is stable if and only if the integral of the absolute value of the impulse function is finite [3.5]. This condition is based on a time-domain analysis. If a feedback system is analyzed in the s-domain (i.e. frequency domain) the position of the transfer function poles in the s-plane is analyzed. It can be shown that the time-domain stability condition corresponds to a frequency-domain stability condition: if one or more transfer function poles lay in the right side of the s-plane, the system will oscillate [3.4]. If all poles lay in the left side of the s-plane the system will be stable (see Figure 3. 12).Poles of a real system may be shifted compared to the simulated system. A pole lying close to the vertical axis could, in reality, be shifted to the right side of the s-plane, and cause system to oscillate. To avoid oscillations systems are designed to have a certain phase and gain margin from the oscillation condition. A feedback system will oscillate if the closed-loop denominator equals zero. For the PLL closed-loop transfer function given in (3.10) we have:

$$|H_0(j\omega_c)/N| = 1 \tag{3.16}$$

$$\arg(H_0(j\omega_c)/N) = -180° \tag{3.17}$$

The solution of Equ. (3.16) is the unity gain angular frequency, ω_c. ω_c is also called the crossover frequency and it represents the PLL bandwidth. The phase margin is defined as $\arg(H_0(j\omega_c)/N) + 180°$ (see Figure 3. 13).

The gain margin is defined as $20\log(|H_0(j\omega)/N|)$ at the frequency where $\arg(H_0(j\omega)/N) = -180°$. The gain margin makes sense only for a system with three poles or more. For one or two pole transfer function the open-loop phase is never $-180°$.

In order to analyse the stability of the type II PLL, with respect to the PLL parameters, let us examine the poles of its transfer function given in Equ. (3.12):

$$s_{1,2} = \frac{I_{cp}K_{VCO}}{4\pi N}R\left(-1 \pm \sqrt{1 - \frac{8\pi NR^2}{I_{cp}K_{VCO}C_p}}\right) \tag{3.18}$$

Figure 3. 14 shows the position of the poles s_1 and s_2 when $K_{VCO}I_{cp}/N$ changes from 0 to ∞. For $K_{VCO}I_{cp}/N = 0$ both poles are at the origin: $s_1 = 0$ and $s_2 = 0$. As $K_{VCO}I_{cp}/N$ grows, the poles move on the circle and reach the real axis for $K_{VCO}I_{cp}/N = 8\pi R_2/C_p$, when $s_1 = s_2 = -\omega_p$ and $\zeta = 1$. For

$K_{VCO}I_{cp}/N \to \infty$, the poles remain on the real axis with $s_1 \to -\omega_p/2$ and $s_2 \to -\infty$. We see that for higher $K_{VCO}I_{cp}/N$ the PLL is more stable.

The stability analysis of a third order PLL (with the ripple suppression capacitor C_1 (Figure 3.11c)) is mathematically much more complex and is given in [3.6]. In short, a third order PLL is more stable for higher $K_{VCO}I_{cp}/N$ and for higher C_2/C_1. When designing a third order PLL, both the phase margin as well as the gain margin have to be observed to make sure that the PLL will not oscillate.

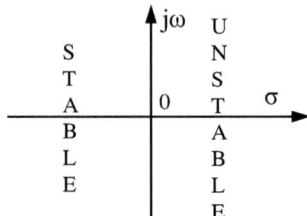

Figure 3. 12 Transfer function pole loci for a stable and an unstable system.

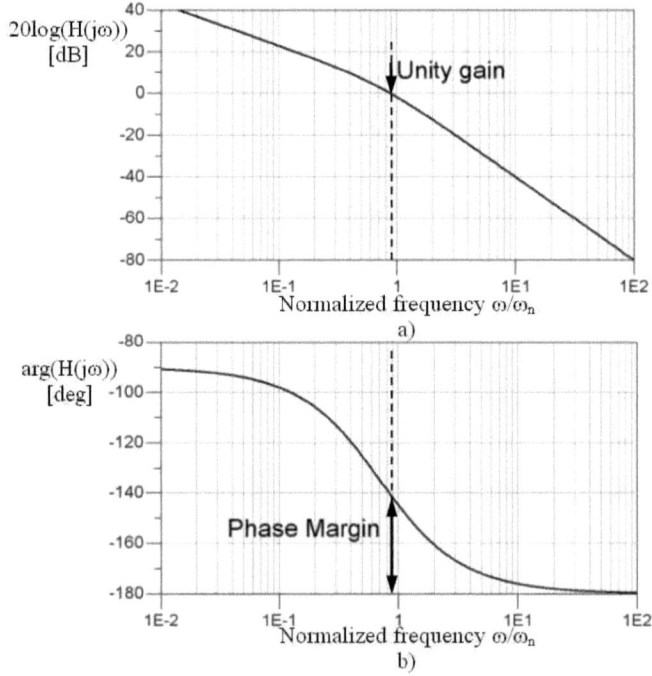

Figure 3. 13 Open loop transfer function phase margin is defined for unity gain frequency shown in a). The margin is shown in b).

3.2.3. PLL Phase Noise Properties

The phase noise level of the PLL output signal is an important performance parameter of a PLL, especially for a PLL which is used for frequency synthesis. The output phase noise comes form the phase noise of the input signal and from the (phase) noise generated in the PLL blocks. Although the noise comes from all PLL blocks, there are three main sources of phase noise in a PLL:

1. Phase noise of the input signal.

2. VCO phase noise, and

3. LPF noise.

To understand the effect of the PLL on each noise source, we have to see the PLL transfer function for each noise type.

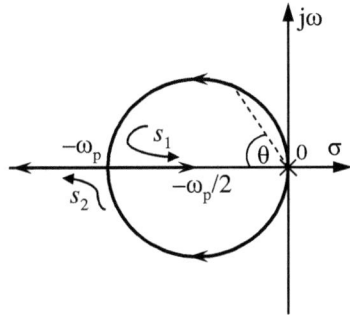

Figure 3. 14 Type II PLL transfer function pole loci.

3.2.3.1. Phase Noise of the Input Signal

Since the input phase noise represents the input excess phase, and the output phase noise the output excess phase, the closed-loop transfer function H(s) represents the PLL transfer function for the input phase noise:

$$\frac{\Phi_{out}}{\Phi_{in}}(s) = H(s) = \frac{H_0(s)}{1 + \frac{H_0(s)}{N}} \tag{3.19}$$

where $H_0(s)$ represents the open-loop transfer function and N represents the division ratio.

The closed-loop transfer function has a shape of a lowpass filter (Figure 3. 15). For $s \to 0$, H(s) $\to$ N. This means that the low frequency phase noise within the PLL bandwidth will pass to the PLL output and be amplified N times. For a high division ratio N, for example N = 1000, low frequency phase noise will be amplified by 60 dB. This implies that PLL frequency synthesizers with high division ratio require low phase noise input reference. For small values of the phase margin ($\Delta\varphi<45°$), the input phase noise is additionally amplified around the crossover frequency (Figure 3.11).

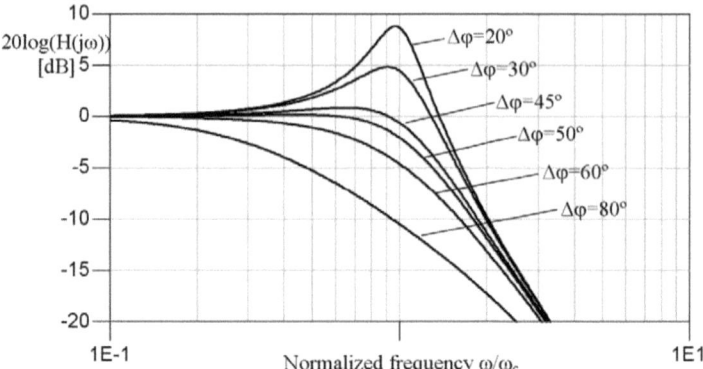

Figure 3.15 PLL input phase transfer function normalized for the crossover frequency, ω_c, for different values of phase margin and division ratio $N = 1$.

3.2.3.2. VCO Phase Noise

The phase noise of the VCO can be modelled as an additive noise after the VCO, $\varphi_{VCO}(t)$ ($\Phi_{VCO}(s)$). This noise is uncorrelated to the input noise $\varphi_{in}(t)$ ($\Phi_{in}(s)$), and if we set the input noise (not the periodic signal) to zero, we can calculate the PLL transfer function for the noise which is added after the VCO. This transfer function is given as [3.4]:

$$\frac{\Phi_{out}}{\Phi_{VCO}}(s) = H_{VCO}(s) = \frac{1}{1 + \frac{H_0(s)}{N}} \quad (3.20)$$

where $H_0(s)$ represents the open-loop transfer function and N represents the division ratio.

Figure 3.16 shows PLL transfer function for the VCO phase noise for different values of phase margin. This is a highpass filter function. It means that the PLL suppresses the VCO noise inside the PLL bandwidth ($\omega < \omega_c$). The VCO phase noise outside the PLL bandwidth is passed thru to the PLL output unsuppressed. For low phase margins, the noise around the crossover frequency will be amplified.

The highpass characteristic of the PLL VCO noise transfer function means that if the VCO phase noise level is high, the PLL will have to be broadband (i.e. high crossover frequency) to reduce the noise level at the PLL output.

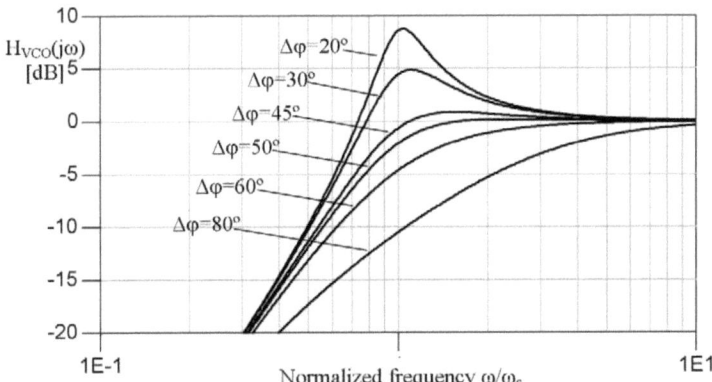

Figure 3.16 PLL VCO phase noise transfer function normalized for the crossover frequency, ω_c, for different values of the PLL phase margin.

3.2.3.3. LPF Noise

The LPF noise is thermal noise that is generated on the LPF resistor(s). This noise is fed to the VCO input and it is transfers into a phase noise at the VCO/PLL output.

Resistor thermal noise can be modelled with a noiseless resistor in a series with a voltage source, which generates voltage $v_n(t)$ with spectral density $V_n(f)$. $V_n(f)$ is given as $V_n(f) = 4kTR$, where k represents Boltzmann's constant (given in J/K), T the absolute temperature (in K) and R the value of the resistor (in Ω). The LPF resistor can have large value (in the order of 1kΩ or 10kΩ) resulting in a significant noise power.

To calculate the level of the LPF noise at the output of the PLL, we have to first calculate the noise voltage at the output of the LPF, and than to calculate the effect of the PLL transfer function on the noise which is added at the LPF output. For a third order PLL with a LPF shown in Figure 3.11c, the LPF output voltage can be calculated as:

$$V_{LPF}(j\omega) = \frac{\frac{1}{j\omega C_1}}{\frac{1}{j\omega C_1} + R_2 + \frac{1}{j\omega C_2}} V_n(\omega) \qquad (3.21)$$

The LPF noise can be modelled as an additive noise, which is uncorrelated to the input noise $\varphi_{in}(t)$ ($\Phi_{in}(s)$), and if we set the input noise to zero, we can calculate the PLL transfer function for the noise which is added after the LPF. This transfer function is given as [3.4]:

$$\frac{\Phi_{out}}{V_{LPF}}(s) = H_{LPF}(s) = \frac{\frac{K_{VCO}}{s}}{1 + \frac{H_0(s)}{N}} \qquad (3.22)$$

where $H_0(s)$ represents the open-loop transfer function and N represents the division ratio and K_{VCO} the gain of the VCO.

Figure 3.17 shows PLL transfer function for the LPF noise for different values of phase margin. This is a bandpass filter function with passband centre frequency equalling the PLL bandwidth frequency ω_c. The contribution of the LPF noise to the PLL output phase noise may be significant if the input and VCO phase noise are relatively low. However, usually it is possible to design a PLL such that the LPF noise contribution at the PLL output is negligible.

3.3. RMS Phase Error Optimization

The bit error rate (BER) of the detection in the receiver baseband depends on several factors. The most important are the phase noise level of the PLL signal and the white noise level. System level simulations have shown that for an OFDM signal with 16QAM modulation, the RMS phase error of the 56/48 GHz PLL signal should be below 3 degrees.

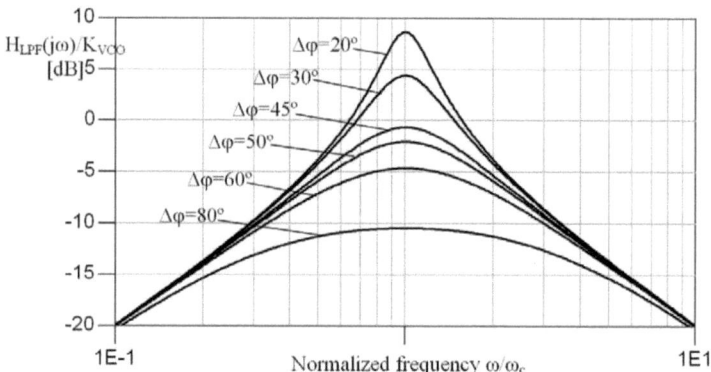

Figure 3.17 PLL LPF phase noise transfer function normalized for the crossover frequency, ω_c, for different values of the PLL phase margin.

The PLL phase noise is a sum of the phase noise form the VCO, LPF and the input reference. Integrating the PLL phase noise gives the RMS phase error, but this value has to be modified due to the common phase error (CPE) correction, which takes place in the baseband. The CPE correction effect can be modelled by filtering the PLL phase noise with a highpass filter with a corner frequency f_c corresponding the symbol duration T_S ($f_c = T_S$). The transfer function of the filter for modelling CPE correction is $H_{CPE}(s) = 1/(1 + j2\pi f_c/s)$.

The exact calculation of the RMS phase error requires knowing all PLL parameters (I_{cp}, K_{VCO}, loop filter elements). It is, however, possible to calculate the RMS phase error with sufficient accuracy by assuming that the main noise contribution comes from the VCO and the input reference phase noise. The PLL loop filter can usually be designed so that the noise contribution from the loop resistors to the total output phase noise is negligible. For a given VCO and input reference phase noise, the RMS phase error can be calculated for different values of the PLL bandwidth and phase margin. The optimum PLL bandwidth and phase margin are chosen for the minimum RMS phase error.

3.4. Calculation of PLL Parameters

Performance of a PLL, as a frequency synthesizer, is determined by its:

1. **Locking range.** The maximum locking range is limited by the oscillation range of the VCO and the minimum and maximum dc control voltage that the CP can sustain at the VCO control input.

2. **Locking time.** PLL locking time depends on the PLL bandwidth and the phase margin. PLL bandwidth is inversely proportional to the locking time (for a fixed phase margin value). The minimum locking time is achieved for the phase margin around 45°.

3. **Output phase noise level.** Depending on the application, the maximum allowed phase noise level may be specified at different frequency offsets from the carrier, or an integrated value of the phase noise may be given (i.e. phase jitter). The output phase noise can be reduced by optimizing the bandwidth and phase margin of the PLL.

4. **Spur level.** For an integer PLL (which is of our interest), the spur depends mainly on charge pump current mismatch and the LPF attenuation at the spur frequencies.

The importance of these parameters depends on the application, i.e. the locking time is not important if the PLL frequency is not going to be switched during the operation. The needed PLL bandwidth and phase margin can be determined based on the required locking time and output phase noise level. These values are used to calculate the parameters of the PLL: I_{cp}, K_{VCO} and loop filter elements (C_1, C_2, R_2). We assume here that the parameters N and K_{VCO} are given, and will not be calculated. They may be chosen to achieve desired frequency range and input frequency.

3.4.1. Parameter Calculation Recipe for a Third Order PLL

The LPF transimpedance can be given as:

$$F(s) = \left(\frac{b}{b+1}\right)\frac{\tau s + 1}{sC_2\left(\frac{\tau s}{b+1}+1\right)} \tag{3.23}$$

where $\tau = R_2C_2$ and $b = C_2/C_1$. The open loop transfer function is than:

$$H_0(s) = \frac{K_{VCO}I_{cp}}{2\pi}\left(\frac{b}{b+1}\right)\frac{\tau s + 1}{s^2 C_2\left(\frac{\tau s}{b+1}+1\right)} \tag{3.24}$$

and the phase margin is given by:

$$\Delta\varphi = \tan^{-1}(\tau\omega_c) - \tan^{-1}\left(\frac{\tau\omega_c}{b+1}\right) \tag{3.25}$$

where ω_c is the or PLL bandwidth. It can be shown by differentiating (3.25) that the maximum phase margin is achieved for $\omega_c = \sqrt{b+1}/\tau$, which gives the maximum phase margin:

$$\Delta\varphi_{max} = \tan^{-1}\left(\sqrt{b+1}\right) - \tan^{-1}\left(\frac{\sqrt{b+1}}{b+1}\right) \tag{3.26}$$

We may notice that the maximum phase margin depends only on b (i.e. ratio of C_2 and C_1).

The crossover frequency is chosen so that it equals the maximum phase margin frequency i.e. $\omega_c = \sqrt{b+1}/\tau$. Then the equation that defines the crossover frequency $|H_0(j\omega_c)/N| = 1$ becomes:

$$\frac{K_{VCO}I_{cp}}{2\pi N}\left(\frac{b}{b+1}\right) = \frac{C_2}{\tau^2}\sqrt{b+1} \tag{3.27}$$

Based on this, the recipe to calculate the PLL parameters is defined as follows [3.7]:

1. Find b from (3.19) based on the required phase margin.

2. For the required PLL bandwidth find τ from $\tau = \sqrt{b+1}/\omega_c$.

3. Chose C_2 and I_{cp} so that they satisfy (3.27).

4. Calculate noise contribution of R_2. If it is acceptable, the design is complete, otherwise go to step 3. and increase C_2.

3.4.2. Parameter Calculation Recipe for a Fourth Order PLL

The fourth order PLL parameter calculation follows the same procedure as for the third order PLL [3.7]. The maximum phase margin is given by:

$$\Delta\varphi_{max} = \tan^{-1}(\tau\omega_c) - \tan^{-1}\left(\frac{A\tau\omega_c}{1 - B(\tau\omega_c)^2}\right) \tag{3.28}$$

where

$$A = \frac{\frac{C_1}{C_2} + \frac{C_3}{C_2} + \frac{\tau_2}{\tau}\left(1 + \frac{C_1}{C_2}\right)}{1 + \frac{C_1}{C_2} + \frac{C_3}{C_2}} \tag{3.29}$$

$$B = \frac{\dfrac{C_1}{C_2}\dfrac{\tau_2}{\tau}}{1 + \dfrac{C_1}{C_2} + \dfrac{C_3}{C_2}}$$

and $\tau_2 = R_3 C_3$. The maximum phase margin is achieved for:

$$\omega_c = \frac{1}{\tau}\sqrt{\frac{1}{2}\left(\frac{2B + AB + A - A^2}{B(B-A)} + \sqrt{\left(\frac{2B + AB + A - A^2}{B(B-A)}\right)^2 - \frac{4(1-A)}{B(B-A)}}\right)} \qquad (3.30)$$

We may notice that the maximum phase margin doesn't depend on the absolute values of LPF elements, but on their ratios (C_1/C_2, C_3/C_2 and R_3/R_2), as was the case for the third order PLL. When the crossover frequency is chosen to equal the maximum phase margin frequency ($f_c/f_{PM} = 1$) the equation that defines the crossover frequency ($|H_0(j\omega_c)|/N| = 1$) becomes:

$$\frac{K_{VCO} I_{cp}}{2\pi}\sqrt{\frac{1 + (\tau\omega_c)^2}{(A\tau\omega_c)^2 + (1 - B(\tau\omega_c)^2)^2}} = C_2\left(1 + \frac{C_1}{C_2} + \frac{C_3}{C_2}\right)\omega_c^2 \qquad (3.31)$$

The recipe to calculate the fourth order PLL parameters is given in literature [3.7], [3.8]. These recipes are analogue to the recipe given for the third order PLL.

When choosing C_1/C_2, C_3/C_2 and R_3/R_2 to achieve the required phase margin, we have two degrees of freedom. This allows us to choose the PLL parameters such to minimize, for example, the spur, while keeping the PLL bandwidth, phase margin and the LPF noise the same.

Spur level at the output of the PLL can be calculated using standard FM modulation theory and the small angle approximation [3.9]. For an integer-N PLL, spurious signals occur at integer multiples of the reference frequency f_{ref}. Amplitude of spurious signals is given in dBc (in dB with reference to the carrier) [3.9]:

$$S_p(f_{ref}) = 20\log\left(\frac{i_{cp}(f_{ref})F(f_{ref})K_{VCO}}{2f_{ref}}\right) \qquad (3.32)$$

where $i_{cp}(f_{ref})$ is amplitude of ac current component at frequency f_{ref}, and $F(f_{ref})$ is the transimpedance of the LPF at f_{ref} [3.9]. Since $i_{cp}(f_{ref})$ is proportional to I_{cp}, spur performance of two PLLs can be compared using:

$$\Delta S_p = 20\log\left(\frac{I_{cp1}|F_1(f_{ref})|K_{VCO1}}{I_{cp2}|F_2(f_{ref})|K_{VCO2}}\right) \qquad (3.33)$$

The spur performance of different PLLs will be compared using this equation and one reference PLL. This is a simpler and a better approach for PLL comparison, than calculating spur level for each PLL.

The modified recipe for minimum spur is presented here:

1. Choose C_1/C_2, C_3/C_2 and R_3/R_2 from (3.28) based on the required phase margin.

For the required PLL bandwidth find τ from (3.30).

2. Chose C_2 and I_{cp} so that they satisfy (3.31).

3. Calculate noise contribution of R_2 and R_3. If it is below the acceptable level go to step 5. If it is higher than the acceptable level go to step 6, and if it equals the acceptable level go to step 7.

4. Increase R_3/R_2. This will reduce the phase margin, so C_3/C_2 has to be increased to achieve the required margin. Find R_3/R_2 for which their noise contribution equals the acceptable level and go to step 7.

5. Reduce R_3/R_2. This will increase the phase margin, so C_3/C_2 has to be reduced to achieve the required margin. Find R_3/R_2 for which their noise contribution equals the acceptable level and go to step 7. If the noise contribution is too high for minimum R_3/R_2 ($R_3/R_2 = 0$, i.e. the PLL becomes a third order PLL) increase C_2 and go to step 3.

6. Find optimum C_1/C_2 for minimal spur (using (3.33)) by following steps 1 to 6 for different values of C_1/C_2.

Let us now apply this recipe to calculate PLL parameters for a 56 GHz PLL with $K_{VCO} = 2$ GHz/V, $N = 512$ (giving $f_{ref} = 109.375$ MHz). The parameters are calculated for PLL bandwidth $fc = 4.2$ MHz, and different phase margin values: 30, 40, 45, 50 and 60° (see Tabel I). The acceptable level of LPF noise at $fc = 4.2$ MHz is −105 dBc/Hz. Spur value for 45° was taken as a referent value. The calculation of parameters was done for constant $C_2 = 50$ pF. It is important to keep the size of the biggest capacitor (C_2) constant to get a fair comparison for integrated PLLs, because larger C_2 increases the area of the PLL layout.

$\Delta\varphi_M$ [°]	C_1 [pF]	C_2 [pF]	C_3 [pF]	R_2 [kΩ]	R_3 [kΩ]	I_{cp} [mA]	S_P [dB]
30	8.75	50	2.01	1.19	4.22	6.2	−4.6
40	6.25	50	0.94	1.53	6.9	4.75	−1.8
45	5	50	0.7	1.75	8.15	4.1	0 (Ref. Value)
50	3.75	50	0.54	2.02	9.48	3.52	1.9
60	1.87	50	0.35	2.78	11.14	2.49	6.6

Table 3. I Calculated PLL Parameters for different phase margins.

Figure 3.18 shows the dependence of spur on phase margin, which is given in Table 3.I. It shows that it is more difficult to make a PLL with higher phase margin. The PLL with 60° margin would have 11.2 dB higher spur than the PLL with 30° phase margin.

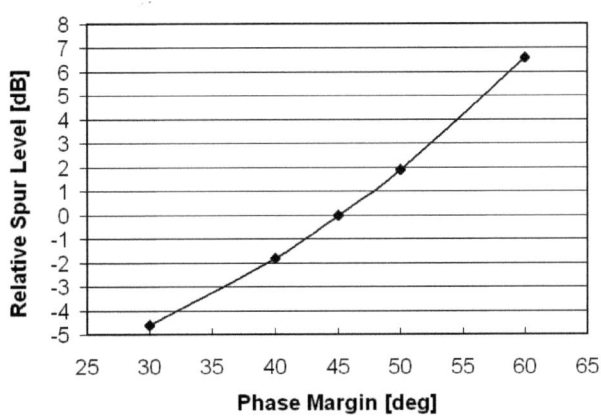

Figure 3.18 Relative value of Spur for different phase margin values and the same PLL bandwidth, LPF noise level and capacitor C_2.

Table 3.II presents PLL parameters for constant phase margin of 45° and different values of PLL bandwidth. K_{VCO}, N and C_2 are the same as for Table 3.I. LPF phase noise is scaled with the bandwidth – for double f_c acceptable LPF noise is 6 dB lower. VCO phase noise drops in the same way so LPF noise level is kept constant relative to the VCO noise.

Figure 3.19. shows the dependence of Spur on PLL bandwidth (given in Table 3.II). We can see that PLL spur performance depends strongly on the PLL bandwidth. The PLL with 8.4 MHz bandwidth would have 32.7 dB higher spur than the PLL with 2.1 MHz bandwidth, meaning that it is much more difficult to make a broadband PLL with good spur performance, than a narrowband one. This is because for a narrowband PLL $\omega_3 = 1/R_3C_3$ can be much lower attenuating the spur more.

f_c [MHz]	C_1 [pF]	C_2 [pF]	C_3 [pF]	R_2 [kΩ]	R_3 [kΩ]	I_{cp} [mA]	PN@f_c [dBc/Hz]	S_P [dB]
2.1	5	50	1.38	3.53	7.05	1.04	−99	−16.3
4.2	5	50	0.7	1.75	8.15	4.1	−105	0 (Ref. Value)
8.4	3.75	50	0.44	0.87	8.96	16	−111	16.4

Table 3.II Calculated PLL Parameters for different PLL bandwidths

3.4.3. A New Parameter Calculation Recipe for a Fourth Order PLL

The parameter calculation recipe presented in the previous section gives PLL parameters where the crossover frequency matches the maximum phase margin frequency ($f_c/f_{PM} = 1$). This approach makes sense for a third order PLL because it gives optimal LPF component values. The capacitor C_2 has minimum value for the acceptable LPF noise. This is important because this capacitor is usually large and for integration it is important to reduce its size.

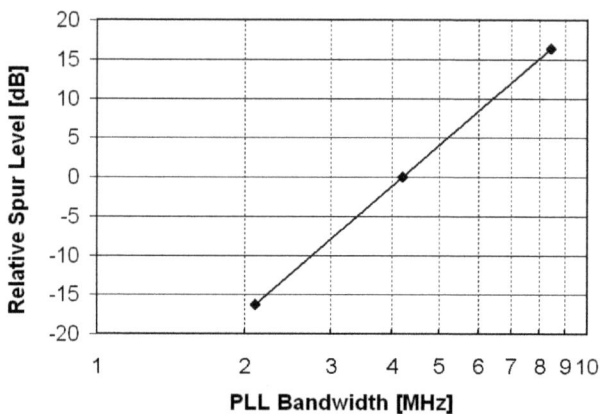

Figure 3.19 Relative value of Spur for different PLL bandwidth values and the same phase margin and capacitor C_2 and scaled LPF noise level.

If we don't limit our self's to equalling the crossover frequency with the maximum phase margin frequency we have one more degree of freedom in determining the parameters. To specify the recipe for parameter calculation for $f_c/f_{PM} \neq 1$ we need two more equations. Phase margin is defined as:

$$\Delta\varphi_{PM} = \arg(H_0(j\omega_c)) \qquad (3.34)$$

The left side of Equ. (3.30) gives the frequency for maximum phase margin. If $X = f_c/f_{PM}$, we can modify (3.23) to:

$$\omega_c = \frac{X}{\tau}\sqrt{\frac{1}{2}\left(\frac{2B+AB+A-A^2}{B(B-A)} + \sqrt{\left(\frac{2B+AB+A-A^2}{B(B-A)}\right)^2 - \frac{4(1-A)}{B(B-A)}}\right)} \qquad (3.35)$$

where A and B are given in (3.29).

The recipe for calculating PLL parameters for minimum spur when $f_c/f_{PM} \neq 1$ is as follows:

1. Choose C_1/C_2, C_3/C_2 and R_3/R_2 from (3.34) based on the required phase margin.

2. Choose f_c/f_{PM} and for the required PLL bandwidth find τ from (3.35).

3. Chose C_2 and I_{cp} so that they satisfy (3.31).

4. Calculate noise contribution of R_2 and R_3. If it is below the acceptable level go to step 5. If it is higher than the acceptable level go to step 6, and if it equals it go to step 7.

5. Increase R_3/R_2. This will reduce the phase margin, so C_3/C_2 has to be increased to achieve the required margin. Find R_3/R_2 for which their noise contribution equals the acceptable level and go to step 7.

6. Reduce R_3/R_2. This will increase the phase margin, so C_3/C_2 has to be reduced to achieve the required margin. Find R_3/R_2 for which their noise contribution equals the acceptable level and go to step 7. If the noise contribution is too high for minimum R_3/R_2 ($R_3/R_2 = 0$, i.e. the PLL becomes a third order PLL) increase C_2 and go to step 3.

7. Find optimum C_1/C_2 for minimal spur (using (3.33)) by following steps 1 to 6 for different values of C_1/C_2.

fc/f_{PM}	C_1 [pF]	C_2 [pF]	C_3 [pF]	R_2 [kΩ]	R_3 [kΩ]	I_{cp} [mA]	S_P [dB]
1	5	50	0.7	1.75	8.15	4.1	0 (Ref. Value)
1.5	3.75	50	0.98	2.85	8.68	2.72	−4.4
2	2.5	50	1.17	4.37	8.51	1.85	−5.7
2.2	2.5	50	1.1	5.09	8.14	1.62	−5.9
2.5	2.5	50	0.99	6.28	7.22	1.34	−5.6
3	1.87	50	1.01	8.55	6.2	1.01	−4.7

Table 3. III Calculated PLL Parameters using old and new technique for different ratios of PLL bandwidth and maximum phase margin frequencies

This recipe was used to calculate PLL parameters for different values of fc/fPM: 1.5, 2, 2.2, 2.5 and 3 (see Table 3. III). The parameters were calculated for: $K_{VCO} = 2$ GHz/V, $N = 512$, $C_2 = 50$ pF and acceptable LPF noise level is −105 dBc/Hz (is for Table I). Figure 3. 20 shows the dependence of spur level on fc/f_{PM} ratio (given in Table 3. III). We see that optimal PLL performance is achieved for $fc/f_{PM} = 2.2$ when PLL spur performance is improved by approximately 6 dB.

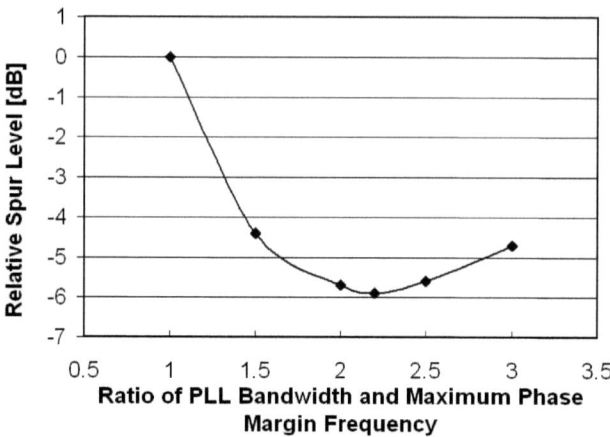

Figure 3. 20 Relative value of Spur for different values of f_c/f_{PM} ratio and the same phase margin, PLL bandwidth, capacitor C_2 and LPF noise level.

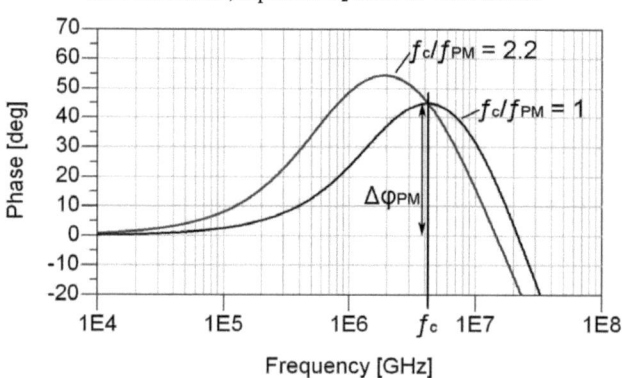

Figure 3. 21 Phase plots for PLL parameters in Table 3. III for $f_c/f_{PM} = 1$ and $f_c/f_{PM} = 2.2$.

f_c/f_{PM}	C_1 [pF]	C_2 [pF]	C_3 [pF]	R_2 [kΩ]	R_3 [kΩ]	I_{cp} [mA]	Improvement
1	5	50	0.7	1.75	8.15	4.1	Sp:0 (Ref. Value)
2.2	2.5	50	1.1	5.09	8.14	1.62	Spur:−5.9
2.2	1	20	1.047	12.58	3.77	0.705	Icp:5.8xSmaller
2.2	3.12	50	2.13	5.03	1.51	1.765	PN:−3.9dBLower

Table 3. IV Calculated PLL Parameters using old and new technique for minimum spur, C_2 capacitor and LPF phase noise.

Figure 3. 21 shows phase plots of open-loop transfer function for the standard approach where $fc/f_{PM} = 1$ (blue line) and the new approach where the crossover frequency is 2.2 times higher than the maximum phase margin frequency ($fc/f_{PM} = 2.2$). Stability gain margin for $fc/f_{PM} = 1$ is approximately 20 dB, and for $fc/f_{PM} = 2.2$ it is app. 14 dB, which is sufficient. These two PLLs were simulated using time-domain behavioural simulation to verify that they have the same settling behaviour. Simulation results for the LPF output voltage (i.e. VCO input control voltage) are shown in Figure 3. 22a. It can be seen that the settling is very similar, with both PLL settling after approximately 400 ns (this of course depends on the settling criteria). It is interesting to see settling of the voltage at the biggest capacitor C_2. For the standard case ($fc/f_{PM} = 1$, red line) the voltage settles together with the output voltage (in approximately 400 ns). For the new case ($fc/f_{PM} = 2.2$, blue line) this voltage takes much more time to settle. This, however, does not have any effect on PLL properties.

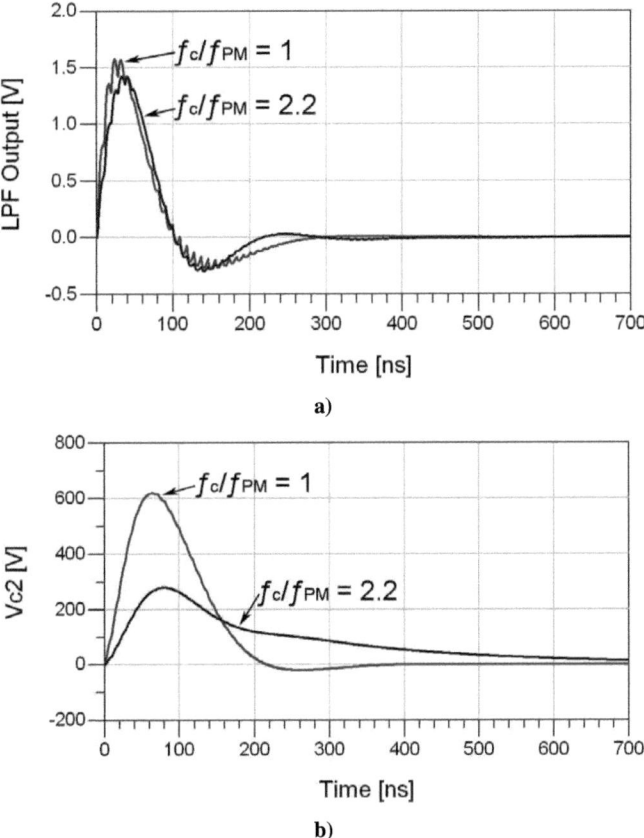

Figure 3. 22 Settling behaviour of PLLs with PLL parameters in Table III for $f_c/f_{PM} = 1$ (red line) and $f_c/f_{PM} = 2.2$ (blue line) (a), settling of the voltage at capacitor C_2 (b).

PLL parameters don't have to optimized for better spur performance. We may modify the recipe in steps 4. 5, 5 and 7 so to choose C_1/C_2, C_3/C_2 and R_3/R_2 for lower LPF noise, or lower charge-pump current I_{cp} (which reduces charge-pump noise). Table 3. IV shows PLL parameters optimized for best spur performance, minimum I_{cp} and minimum capacitor C_2, and minimum LPF noise.

fc/f_{PM}	C_1 [pF]	C_2 [pF]	C_3 [pF]	R_2 [kΩ]	R_3 [kΩ]	I_{cp} [mA]	S_P [dB]
1	2	20	0.68	4.42	6.52	1.675	0 (Ref. Value)
2.2	1	20	1.04	12.6	3.77	0.705	−1.6
1	5	50	0.7	1.75	8.15	4.1	−1.7
2.2	2.5	50	1.1	5.09	8.14	1.62	−7.6
1	7.5	75	0.7	1.17	8.45	6.15	−2.1
2.2	3.75	75	1.12	3.4	9.11	2.38	−8.8
1	10	100	0.69	0.87	8.74	8.15	−2.3
2.2	5	100	1.13	2.56	9.59	3.14	−9.3

Table 3. V Calculated PLL Parameters using old and new technique for different ratios of PLL bandwidth and maximum phase margin frequencies.

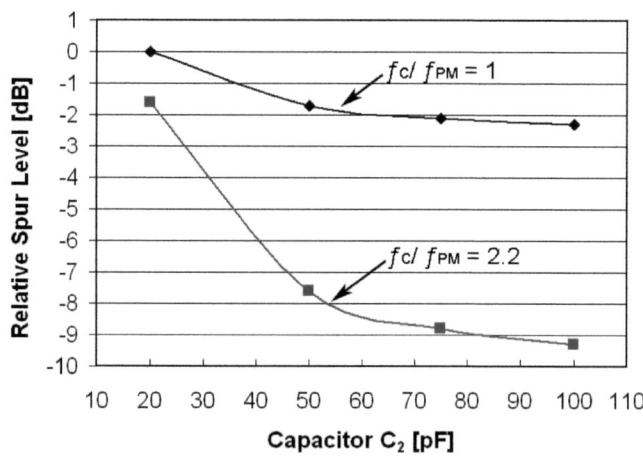

Figure 3. 23 Relative spur level dependency on the size of capacitor C_2 for $f_c/f_{PM} = 1$ and $f_c/f_{PM} = 2.2$.

The improvement of PLL spur performance depends on the size of capacitor C_2. PLL parameters were calculated for different values of C_2 (see Table 3. V). $C_2 = 20$ pF is the minimum size of C_2 for which LPF noise has acceptable level. Figure 3. 23 shows improvement of spur attenuation for $fc/f_{PM} = 1$ and for $fc/f_{PM} = 2.2$. We see that increasing capacitor C_2 reduces spur by a relatively small amount – up to 2.3 dB. Appling the new technique for PLL parameters reduces the spur by up to almost 10 dB.

The new technique is better than the standard one, because by making $f_{PM} < f_c$, $\omega_3 = 1/R_3C_3$ can be lower attenuating the spur more. If $f_{PM} \ll f_c$, phase margin is much lower that the phase maximum, and it is difficult to achieve required phase margin.

3.5. Comparison of Different PLL Topologies

3.5.1. Dual Loop PLL

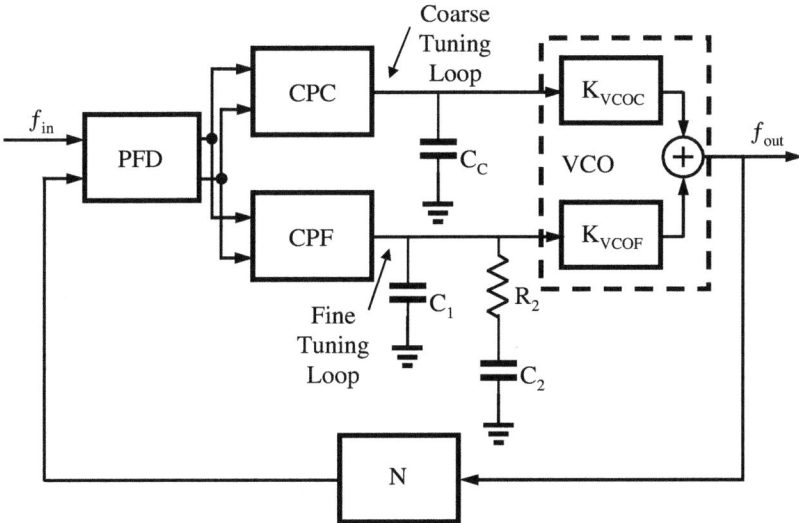

Figure 3. 24 Topology of a dual-loop PLL.

Figure 3. 24 presents topology of a dual loop PLL [3.10]. It has two loops: the fine and coarse loop. There are two charge pumps with currents I_{cpF} and I_{cpC}. The VCO has two varactors, and hence two inputs: V_{ctrlF} and V_{ctrlC}. Total K_{VCO} equals $K_{VCOF} + K_{VCOC}$. Since $\omega_{out} = \omega_0 + K_{VCOF}V_{ctrlF} + K_{VCOC}V_{ctrlC}$ (see Equ. (3.4)), we can introduce effective control voltage $V_{ctrlEFF}$:

$$V_{ctrlEFF} = \frac{K_{VCOF}}{K_{VCO}} V_{ctrlF} + \frac{K_{VCOC}}{K_{VCO}} V_{ctrlC} \qquad (3.36)$$

We can now treat the dual loop PLL as a PLL with one effective loop. The open-loop transfer function is then:

$$H_0(s) = \frac{I_{cpF}}{2\pi} F_F(s) \frac{K_{VCOF}}{s} + \frac{I_{cpC}}{2\pi} F_C(s) \frac{K_{VCOC}}{s} \qquad (3.37)$$

where $F_F(s) = (1/sC_1)\|(R_2 + 1/sC_2)$ and $F_C(s) = 1/sC_C$. If the fine loop part in Equ. (3.30) is much larger than the coarse loop part, the PLL dynamic behaviour will be determined by the fine loop parameters. It can be shown that this condition is satisfied when:

$$\frac{I_{cpF} K_{VCOF} C_C}{I_{cpC} K_{VCOC} C_2} \gg 1 \qquad (3.38)$$

Calculation of PLL parameters for a dual loop can be done using standard procedure for the fine loop parameters (given in section 3.3.1), than calculating $K_{VCOC} = K_{VCO} - K_{VCOF}$ and choosing I_{cpC} and C_C so that they satisfy Equ. (3.38).

We should note here that the filter in the fine loop can be chosen differently. It can be a third order filter (see Figure 3. 11d) for better spur suppression. In that case fine loop PLL parameters can be calculated using procedure given in section 3.3.3. Then we calculate $K_{VCOC} = K_{VCO} - K_{VCOF}$ and choose I_{cpC} and C_C so that the fine loop part in Equ. (3.37) is much larger than the coarse loop part. It can be shown that with a third order filter in the fine loop this condition is satisfied with Equ. (3.38).

3.5.2. Comparison of a Third, Forth Order and Dual Loop PLLs

Table 3. VI presents calculated PLL parameters for a third order, fourth order PLL and for dual loop PLL (with second order filter in the fine loop). All parameters were calculated for K_{VCO} = 2 GHz/V and N = 512. The parameters were calculated for a broadband PLL (fc = 4.2 MHz) and for a narrowband PLL (fc = 0.42 MHz). Phase margin is 45° in both cases. Acceptable LPF noise is −105 dBc/Hz at 4.2 MHz offset and −85 dBc/Hz at 0.42 MHz offset.

For the broadband case (fc = 4.2 MHz), the forth order PLL gives spur level (Sp in the table) of 0 dB (referent value for spur performance comparison), using C_2 capacitor size 50 pF. The third order PLL achieves the same LPF noise performance for much smaller C_2 capacitor of just 6 pF, but it has higher spur level by 12.6 dB. If we use C_2 = 50 pF, LPF noise will drop to −114 dBc/Hz and the spur level will stay the same. Dual loop parameters were calculated for two cases: K_{VCOC} = (1/9)K_{VCOF} and K_{VCOC} = 9K_{VCOF}. Capacitor size is $C_2 + C_C$ = 50 pF. In the first case (K_{VCOC} = (1/9)K_{VCOF}) LPF noise is −105 dBc/Hz and spur level is 13.7 dB higher than for the forth order PLL. For K_{VCOC} = 9K_{VCOF}. LPF noise is much lower: −124 dBc/Hz, but the spur level is much higher: 32.8 dB. We can see in the table that I_{cpF}/I_{cpC} ratio was changed to satisfy condition in Equ. (3.38). We see that dual loop PLL enables trade-off between PLL LPF noise performance and spur performance. Reducing LPF noise level increases spur level, and vice versa.

For the narrowband case ($fc = 0.42$ MHz) the forth order PLL gives high LPF noise. C_2 capacitor was increased to 150 pF to reduce the noise to the acceptable level of –85 dBc/Hz at 0.42 MHz offset. Spur performance is very good with –41 dB lower spur compared to the broadband case. For the third order PLL capacitor C_2 was also increased to achieve acceptable LPF noise level, but only to 60 pF (compared to 150 pF for the forth order PLL). The spur level is –27.4 dB. Dual loop parameters were calculated for two cases: $K_{VCOC} = (71/29)K_{VCOF}$ and $K_{VCOC} = 9K_{VCOF}$. Capacitor size didn't have to be changed: $C_2 + C_C = 50$ pF. In the first case ($K_{VCOC} = (71/29)K_{VCOF}$) LPF noise is –85 dBc/Hz and spur level is –16 dB. For $K_{VCOC} = 9K_{VCOF}$ LPF noise is reduced to –94 dBc/Hz and spur level is increased to –7.2 dB.

PLL Type	f_c [MHz]	PN@f_c [dBc/Hz]	C_1 [pF]	C_2 [pF]	C_3/C_C [pF]	R_2 [kΩ]	K_{VCOF}/K_{VCOC}	R_3 [kΩ]	I_{cpF} (I_{cpC}) [μA]	S_P [dB]
4th order	4.2	–105	2.5	50	1.1	5.09	–	8.14	1620	0 Ref.
3rd order	4.2	–105	1.243	6	–	15.25	–	–	536	12.6
3rd order	4.2	–114	10.36	50	–	1.83	–	–	4470	12.6
Dual Loop	4.2	–105	1.035	5	45	18.3	9/1	–	496 (496)	13.7
Dual Loop	4.2	–124	1.035	5	45	18.3	1/9	–	4480 (56)	32.8
4th order	0.42	–85	7.5	150	10.23	16.7	–	1.17	55	–41
3rd order	0.42	–85	12.43	60	–	15.25	–	–	53.6	–27.4
Dual Loop	0.42	–85	1.035	5	45	183	29/71	–	15.4 (0.77)	–16
Dual Loop	0.42	–94	1.035	5	45	183	1/9	–	44.8 (1.12)	–7.2

Table 3. VI Calculated PLL Parameters for Different PLL Topologies.

Based on the presented comparison we can see what PLL topology should be used based on the required PLL bandwidth, phase margin, spur level, acceptable LPF noise level and PLL size (i.e. $C_2(+C_C)$ size for an integrated PLL):

We should use the simplest topology, i.e. third order PLL. If this topology gives too high spur level (more likely to be the case for a broadband PLL) we should use forth order PLL. If the LPF noise is too high, we should use dual loop PLL. If the third order PLL gives both too high LPF noise and spur level we should use dual loop PLL with third order filter in the fine loop. We may also decide to use forth order or dual loop PLL because some PLL parameters are easier to realize.

3.6. PLL for AFE version I

The PLL for the AFE version I was first produced and tested as a separate chip, and then integrated in the RX and TX. The first version was produced with PLL bandwidth of 4.2 MHz and phase margin 38° [3.11]. Division ratio is 256 (eight cascaded divide-by-two dividers). This PLL features forth order topology, which is well suited for a broadband PLL. The PLL design is based on a PLL that was designed by a colleague and presented in [3.12]. The PLL in [3.12] is a narrowband and features a dual loop topology.

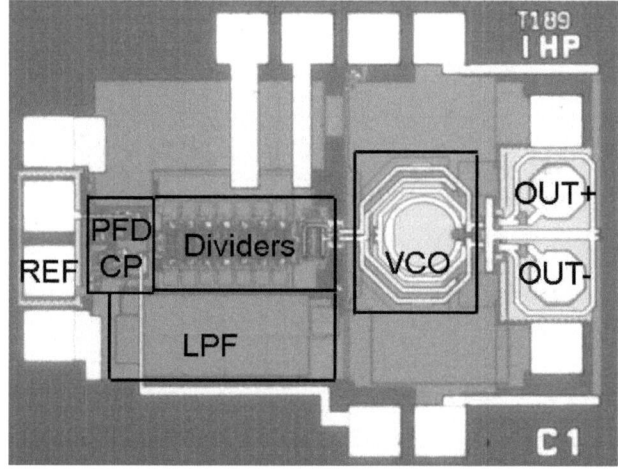

Figure 3. 25 56 GHz PLL micrograph.

PLL parameters are $C_1 = 1$ pF, $C_2 = 35$ pF, $R_2 = 6$ kΩ, $C_3 = 2$ pF, $R_3 = 8$ kΩ and $I_{cp} = 0.75$ mA. VCO features differential Colpitts topology and oscillates in the range of 3.5 GHz around 56 GHz [3.12]. VCO and dividers use HBT transistors, and PFD and CP use CMOS transistors.

Chip photo is shown in Figure 3. 25. Chip size is 900×700 µm^2 including pads and 650×500 µm^2 excluding pads. The PLL requires 3 V and 4.3 V supply. It consumes around 600 mW. The measurements were performed on wafer using the R&S spectrum analyzer FSEM 30 in conjunction with the harmonic diode mixer FS-Z75 for frequency extension to the V-band. To prevent mixer overload, a 20 dB wave-guide attenuator is inserted. The supply voltage is 3 V except for the first divide-by-two stage, which needs 4.3 V. A signal generator provides a sine-wave reference signal from 215.4 to 229.3 MHz.

The measured bandwidth is around 4.5 MHz (Figure 3. 26). The measured phase noise inside the PLL bandwidth is high (−82 dBc/Hz) because the reference signal generator phase noise is too high and it is amplified inside the PLL bandwidth by 48 dB (20log256). This is not a problem because the integrated PLL uses quartz oscillator with low phase noise as a reference. The spur level is below −55 dBc for the whole locking range and as low as −64 dBc (Figure 3. 27).

The PLL that was integrated in the RX and TX was designed for somewhat different PLL bandwidth and phase margin. The parameters were recalculated because the quartz oscillator had higher phase noise floor than previously expected. PLL division ratio was changed to 512 (9 dividers), because quartz oscillators around 100 MHz are cheaper and offer better phase noise floor level. PLL bandwidth was reduced to 3.1 GHz and phase margin was changed to 45°, for lower total phase noise. PLL parameters for this PLL are $C_1 = 3$ pF, $C_2 = 60$ pF, $R_2 = 5.6$ kΩ, $C_3 = 1.4$ pF, $R_3 = 8.7$ kΩ and $I_{cp} = 1.08$ mA. The PLL requires 3 V and 2.6 V supply. It consumes around 400 mW. PLL size with pads is 900×700 μm².

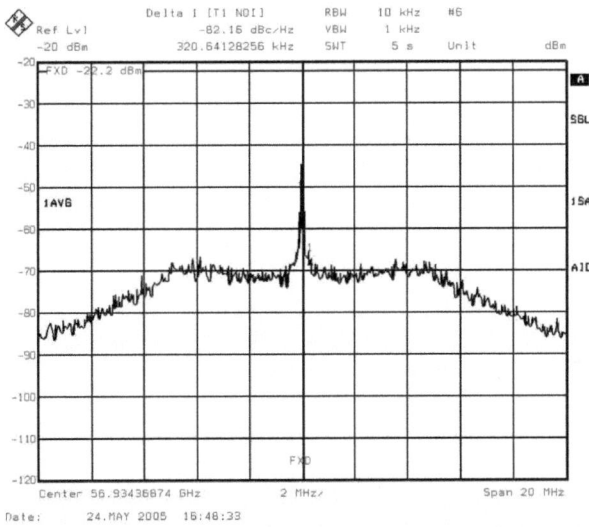

Figure 3. 26 Close spectrum (20 MHz) of the PLL output signal.

Figure 3. 28 shows phase noise of a 100 MHz baseband signal, which was upconverted to 61.35GHz, transmitted, received and downconverted to 100 MHz. The phase noise of that signal is a sum of phase noise of the original signal (negligible), IF PLL phase noise (dominates for lower frequency offsets, below 500 kHz) and 56 GHz PLL phase noise (dominates for higher frequency offsets, above 500 kHz). Phase noise level at 1 MHz offset is −91 dBc/Hz, which means that one 56 GHz PLL has phase noise of −94 dBc/Hz at the same offset. The phase noise at this offset is mainly contributed by the amplified phase noise from the quartz reference oscillator. It is −150 dBc/Hz and after amplification in the PLL (by 20log512=54 dB) it is −96 dBc/Hz. The used reference quartz oscillators had somewhat higher phase noise than was planned (−150 dBc/Hz instead of −155 dBc/Hz), but this was not a limiting factor for the AFE performance.

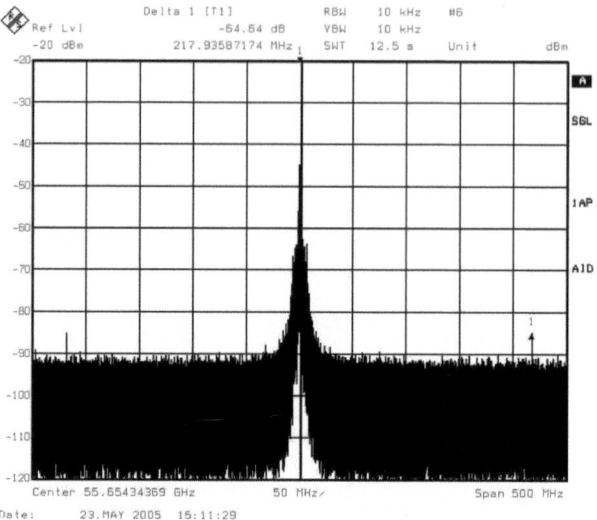

Figure 3. 27 Broad spectrum (500 MHz) of the PLL output signal.

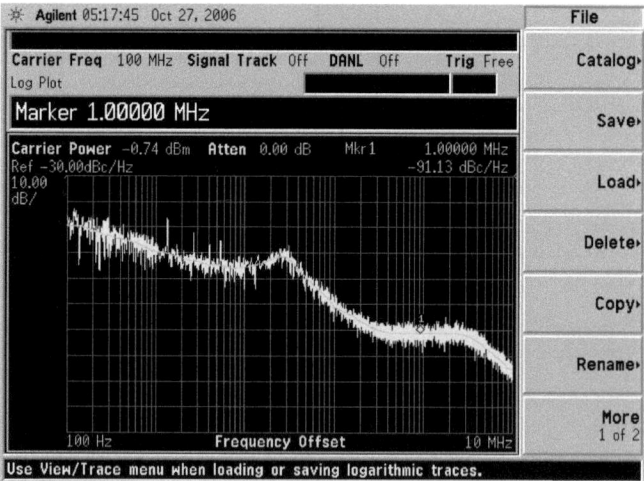

Figure 3. 28 Phase noise spectrum of a 100 MHz baseband signal.

We can see on the picture that the bandwidth matches well with the calculation (around 3.1 MHz). Spurious signals were not seen on the spectrum analyzer (because the signal level is low), but they should be below −45 dBc, according to the simulation.

3.7. PLL for AFE version II

The PLL for AFE version II was presented in [3.13] and was designed by a colleague. It features the sliding-IF topology. This means that in addition to the PLL output from the VCO (at 48 GHz), the PLL has I and Q IF frequency outputs (at 12 GHz). The IF outputs are generated by dividing-by-four the VCO signal.

The measured VCO phase noise (in a stand-alone PLL) is −98 dBc/Hz at 1 MHz offset. This is a very good (i.e. low) value for this frequency range. The RMS phase error from the VCO phase noise is low (below 3 degrees). The PLL was, hence, designed as a narrowband, because the VCO phase noise doesn't have to be attenuated by the PLL. The PLL features a dual-loop topology, which is easy to realize for a narrowband PLL.

The PLL micrograph is shown in Figure 3. 29. PLL size with pads is 1×1.2 mm². It consumes 290 mW from a 3.3 V supply. The measured PLL bandwidth is 150 kHz, and the tuning range is from 47.2 to 49.6 GHz.

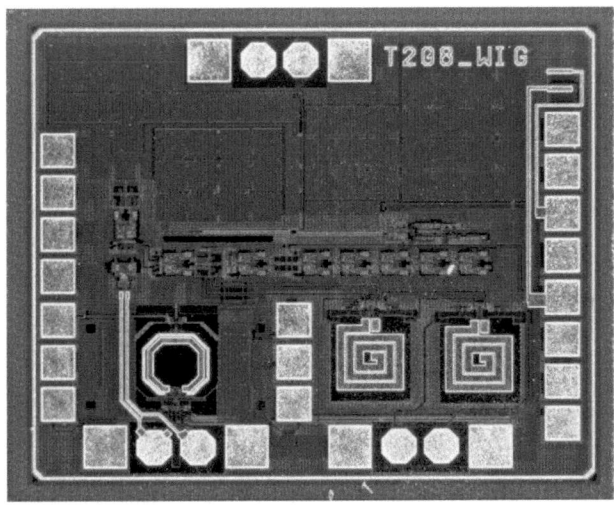

Figure 3. 29 48 GHz PLL micrograph.

3.8. Summary

In this chapter we presented basic PLL theory, and different PLL topologies that can be used: third order, forth order PLL and dual loop PLL. Recipes for calculation of PLL parameters were presented, including new optimized recipe for calculating PLL parameters of a forth order PLL. It was shown that by choosing $f_c/f_{PM} = 2.2$ (ratio of crossover frequency and maximum phase margin frequency) we can design PLL with better spur performance, lower LPF noise, smaller charge pump current or smaller capacitor size (for area reduction of integrated PLLs). Using this approach we can reduce spur by 6 dB, and up to 10 dB.

We have also analyzed different PLL topologies, and concluded that if the simplest, third order PLL topology doesn't satisfy the requirement, we should use the forth order topology for lower spur, dual loop for lower LPF noise and dual loop with third order filter in the fine loop for reducing both the spur level and LPF noise.

The PLLs designed for the AFE version I and II were presented, together with the measurement result.

Chapter 4

Image–rejection Filter

4.1. Introduction

In this chapter, the design of image–rejection filters for 60 GHz applications will be addressed. The goal is analysis, design and optimization of integrated image–rejection filters for 60 GHz. The standard filter design procedure for low–pass filter prototypes and transformations used to convert a lowpass topology to bandpass topology, which is of interest for the intended 60 GHz AFEs. Microstrip bandpass filter structures such as end–coupled, parallel–coupled, hairpin, trisection, quadruplet as well as lumped element filters were examined. Two integrated and one on–board image–rejection filters for 60 GHz signals with image at 50 GHz were designed and produced. Two more integrated filters for the IEEE 802.15.3c standard (image below 40 GHz) were designed and produced. The results are compared with the state–of–the–art.

4.2. Filter Design Theory

The theory on filter design is both old and well established. Physical structures used for realization of filters are very diverse, but their network topology presentation is common. In this section part of the filter design theory, which is relevant for the design of 60 GHz image–rejection filters is presented.

4.2.1. Two–Port Network Characterization

Figure 4.1 shows a two–port network, which can be used to represent most filters. V_1 and V_2 are voltage variables at the ports 1 and 2, respectively. I_1 and I_2 are current variables at the ports 1 and 2, respectively. Z_S is the source or generator impedance and V_S is source or generator voltage. Z_L is load impedance.

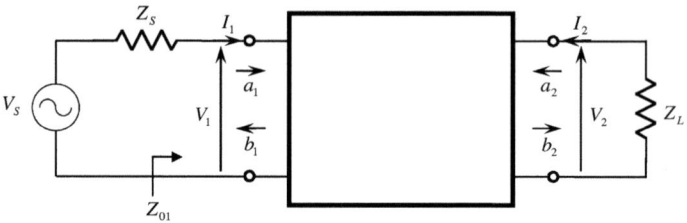

Figure 4.1 Two–port network showing network variables.

The measurement of voltages and currents at microwave frequencies is difficult, so, the incident a_1 and a_2 and the reflected b_1 and b_2 waves were introduced, for ports 1 and 2, respectively. The relationship between voltages and currents and incident and reflected waves is given by:

$$V_n = \sqrt{Z_{0n}}(a_n + b_n) \qquad n = 1,2$$
$$I_n = \frac{1}{\sqrt{Z_{0n}}}(a_n - b_n) \qquad n = 1,2 \qquad (4.1)$$

or

$$a_n = \frac{1}{2}\left(\frac{V_n}{\sqrt{Z_{0n}}} + \sqrt{Z_{0n}}I_n\right) \qquad n = 1,2$$
$$b_n = \frac{1}{2}\left(\frac{V_n}{\sqrt{Z_{0n}}} - \sqrt{Z_{0n}}I_n\right) \qquad n = 1,2 \qquad (4.2)$$

where Z_{0n} denotes impedance at port n.

The scattering or S parameters are used for measurement and characterization of filters and other networks at microwave frequencies. They are defined on the incident and reflected wave variables:

$$S_{11} = \left.\frac{b_1}{a_1}\right|_{a_2=0}$$
$$S_{12} = \left.\frac{b_1}{a_2}\right|_{a_1=0} \qquad (4.3)$$

$$S_{21} = \left.\frac{b_2}{a_1}\right|_{a_2=0}$$

$$S_{22} = \left.\frac{b_2}{a_2}\right|_{a_1=0}$$

where $a_n = 0$ means no reflection at port n (perfect matching).

S parameters are in general complex and may be expressed in terms of amplitude and phase, i.e. $S_{mn} = |S_{mn}|e^{j\arg(S_{mn})}$. for m, $n = 1,2$. S parameters are often presented in decibels $S_{mn}[dB] = 20\log(|S_{mn}|)$ for $m, n = 1,2$.

4.2.2. Filter Parameters

There are three parameters which are commonly used to describe filter performance: the insertion loss, the return loss and the group delay. They are easily defined in terms of S parameters. The insertion loss represents lost energy which does not arrive from the input port m to the termination on the output port n, i.e.:

$$L_A = -20\log(|S_{mn}|)[dB] \qquad\qquad m,n = 1,2 (m \neq n) \qquad (4.4)$$

Return loss represents the portion of the energy which is reflected at the input port n:

$$L_R = -20\log(|S_{nn}|)[dB] \qquad\qquad n = 1,2 \qquad (4.5)$$

The group delay represents the signal (baseband signal, not carrier) delay from the input port to the output port. This parameter is also called the envelope delay. It is defined as:

$$\tau_d = -\frac{d\arg(S_{21})}{d\omega}[s] \qquad (4.6)$$

where port 2 is the output port and port 1 the input port, and ω is the angular frequency.

Before designing a filter, target filter parameters have to first be specified. For a lowpass filter those are: ω_p – passband or cutoff edge angular frequency; A_{max} – maximum insertion loss in the passband; ω_s – stopband edge angular frequency; A_{min} – minimum insertion loss in the stopband, and source and load impedance – which is typically 50 Ohms.

The parameters for a bandpass filter are defined as : ω_{p1} – passband lower edge angular frequency; ω_{p2} – passband upper edge angular frequency; $\omega_0 = \sqrt{\omega_{p1}\omega_{p2}}$ passband centre angular frequency; $B = \omega_{p2} - \omega_{p1}$ filter bandwidth; A_{max} – maximum insertion loss in the passband; ω_{s1} – lower stopband edge angular frequency; ω_{s2} – upper stopband edge angular frequency; A_{min} – minimum insertion loss in the stopbands, source and load impedance. These parameters are shown in Figure 4.2.

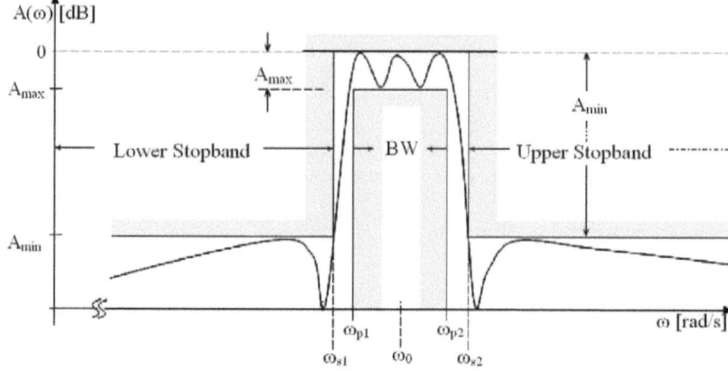

Figure 4.2 A bandpass filter response with specifications.

4.2.3. Typical Filter Response Approximations

A mathematical expression of S_{21} parameter of a two–port filter is called the transfer function or the frequency response of the filter ($T(j\omega) = S_{21}(j\omega)$). There are several typical functions which are used for filter response approximation. Those are: Butterworth, Chebyshev, Elliptic (Cauer), Gaussian and All–pass response.

Butterworth response is known as maximally flat response. It has no ripple – neither in the passband nor in the stopband. The magnitude of the transfer function of an n-th order Butterworth filter is given by:

$$|T(j\omega)| = \frac{1}{\sqrt{1+(\omega/\omega_p)^{2n}}} \qquad (4.7)$$

where ω_p is the filter passband edge angular frequency. The insertion loss at the cutoff frequency for Butterworth filter is 3.01 dB. Figure 4.3 shows Butterworth response for orders $n = 1, 3$ and 10.

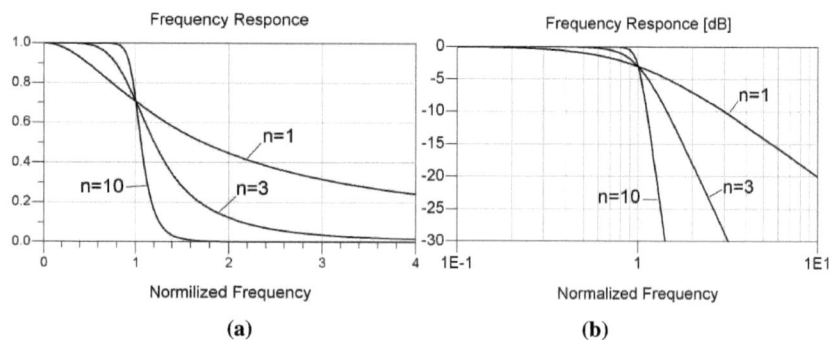

Figure 4.3 A lowpass Butterworth filter frequency response (n=1,3,10)(a), and in dB(b).

Chebyshev response has ripples in the passband, and no ripples in the stopband. The transfer function of a *n*-th order Butterworth filter is given by:

$$|T(j\omega)| = \frac{1}{\sqrt{1 + \varepsilon^2 T_n^2(\omega/\omega_p)}} \tag{4.8}$$

where ω_p is the filter passband edge angular frequency, ε the ripple constant ($\varepsilon = \sqrt{10^{A_{max}/10} - 1}$), and $T_n(\omega)$ is the *n*-th order Chebyshev polynomial of the first kind, defined by:

$$T_n(\omega) = \begin{cases} \cos(n\cos^{-1}\omega), & |\omega| \le 1 \\ \cosh(n\cosh^{-1}\omega), & |\omega| > 1 \end{cases} \tag{4.9}$$

Chebyshev filter is more selective then the Butterworth filter, but it has ripples in the passband and greater group delay.

Elliptic or Cauer response has ripples both in the passband and in the stopband. These filters are the most selective of all filters.

Gaussian response has maximally flat group delay.

All–pass response has unity amplitude response at all frequencies. There is, however, phase shift and group delay, which is frequency dependent.

4.2.4. Lowpass Prototype Filters

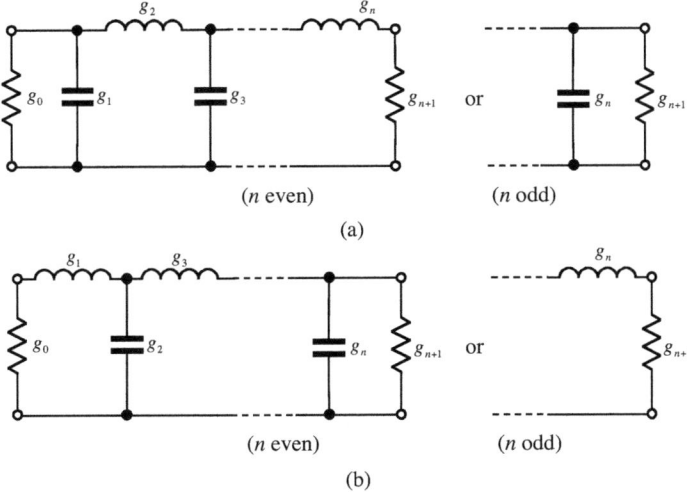

Figure 4.4 Lowpass filter prototype (a) and its dual (b) for all–pole filters with ladder network structures.

Synthesizing a filter with a filter response function, such as those presented in the previous section, is usually done using lowpass prototype filters (show in Figure 4.4) [4.1–2]. The elements of these filters are normalized. The impedance or admittance is normalized with the source and load resistance or admittance. This results in unity value for the source and load impedance or admittance, i.e. $g_0 = 1$, $g_{n+1} = 1$. The angular frequency is normalized with the cutoff angular frequency giving: $\Omega = \omega/\omega_p = 1$.

Butterworth lowpass prototype filter with insertion loss $L_A = 3.01$ dB at normalized angular cutoff frequency $\Omega_c = 1$, can be computed as follows:

1. Filter parameters have to be defined: ω_p, ω_s (cutoff and stopband edge angular frequency) A_{max}, A_{min} (maximum and minimum insertion loss).

2. Minimum required filter order can be calculated by:

$$n \geq \frac{\log\left(\frac{10^{\frac{A_{min}}{10}} - 1}{10^{\frac{A_{max}}{10}} - 1}\right)}{2\log(\omega_p/\omega_s)} \quad (4.10)$$

3. Prototype elements can be calculated by:

$$g_0 = 1$$

$$g_i = 2\sin\left(\frac{(2i-1)\pi}{2n}\right) \qquad i = 1, 2, \ldots, n \quad (4.11)$$

$$g_{n+1} = 1$$

Chebyshev lowpass prototype filter with insertion loss L_A at normalized angular cutoff frequency $\Omega_c = 1$, can be computed as follows in a similar way as the Butterworth filter:

1. Filter parameters have to be defined: ω_p – cutoff edge angular frequency; A_{max} maximum insertion loss in the passband; ω_s – stopband edge angular frequency; A_{min} minimum insertion loss in the stopband.

2. The ripple constant has to be calculated $\varepsilon = \sqrt{10^{A_{max}/10} - 1}$.

3. Minimum required filter order is calculated by:

$$n \geq \frac{\cosh^{-1}\sqrt{\frac{10^{\frac{A_{min}}{10}} - 1}{10^{\frac{A_{max}}{10}} - 1}}}{\cosh^{-1}\Omega_s} \quad (4.12)$$

where $\Omega_s = \omega_s/\omega_p$.

4. Chebyshev prototype elements are then calculated by:

$$g_0 = 1$$

$$g_1 = \frac{2}{\gamma}\sin\left(\frac{\pi}{2n}\right)$$

$$g_i = \frac{1}{g_{i-1}}\frac{4\sin\left[\frac{(2i-1)\pi}{2n}\right]\sin\left[\frac{(2i-3)\pi}{2n}\right]}{\gamma^2 + \sin^2\left[\frac{(i-1)\pi}{n}\right]} \qquad i = 1,2,\ldots,n \qquad (4.13)$$

$$g_{n+1} = \begin{cases} 1.0 & \text{for } n \text{ odd} \\ \coth^2\left(\frac{\beta}{4}\right) & \text{for } n \text{ even} \end{cases}$$

where

$$\beta = \ln\left[\coth\left(\frac{A_{max}}{17.37}\right)\right] \qquad (4.14)$$

$$\gamma = \sinh\left(\frac{\beta}{2n}\right)$$

4.2.5. Element and Frequency Transformation

Filter prototypes are designed for normalized frequency and element values. To get the "real" filter element values both impedance (impedance scaling) and frequency (frequency mapping) have to transformed. If we denote the source impedance as Z_0 ($Y_0 = 1/Z_0$) the scaling factor γ_0 is:

$$\gamma_0 = \begin{cases} Z_0/g_0 & \text{if } g_0 \text{ represents the resistance} \\ g_0/Y_0 & \text{if } g_0 \text{ represents the conductance} \end{cases} \qquad (4.15)$$

The impedance scaling is applied in the following fashion:

$$g_i \rightarrow \gamma_0 g_i \qquad \text{if } g_i \text{ represents the impedance (resistance or inductance)}$$

$$g_i \rightarrow \frac{g_i}{\gamma_0} \qquad \text{if } g_i \text{ is the admittance (conductance or capacitance)} \qquad (4.16)$$

Frequency mapping allows us to transform a lowpass prototype filter to a practical lowpass, highpass, bandpass or bandstop filter.

Lowpass transformation for a filter with cutoff angular frequency ω_c is given by:

$$\Omega = \left(\frac{\Omega_c}{\omega_c}\right)\omega \qquad (4.17)$$

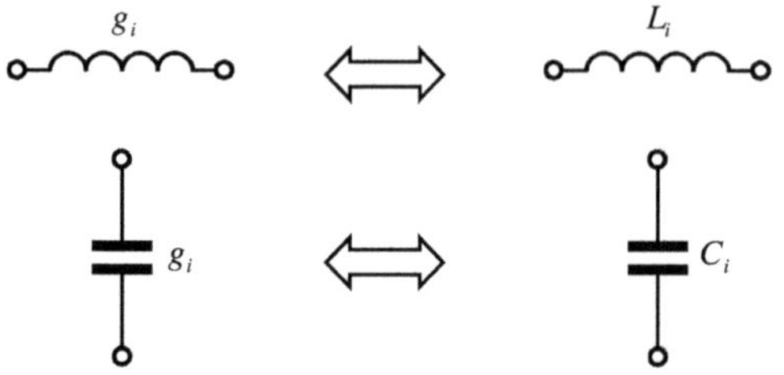

Figure 4.5 Transformation of lowpass prototype filter elements to practical lowpass filter.

If this transformation is applied together with the impedance scaling, element transformation will be:

$$L_i = \left(\frac{\Omega_c}{\omega_c}\right)\gamma_0 g_i \qquad \text{if } g_i \text{ represents the inductance}$$

$$C_i = \left(\frac{\Omega_c}{\omega_c}\right)\frac{g_i}{\gamma_0} \qquad \text{if } g_i \text{ is the capacitance} \qquad (4.18)$$

Bandpass transformation for a filter with ω_{p1} and ω_{p2} (passband lower and upper edge angular frequency), the transformation is given by:

$$\Omega = \frac{\Omega_c}{\Delta}\left(\frac{\omega}{\omega_0} - \frac{\omega_0}{\omega}\right) \qquad (4.19)$$

where $\omega_0 = \sqrt{\omega_{p1}\omega_{p2}}$ is passband centre angular frequency, and fractional bandwidth Δ is:

$$\Delta = \frac{\omega_{p2} - \omega_{p1}}{\omega_0} \qquad (4.20)$$

If this transformation is performed on a prototype element g_i, we will see that this operation transform an inductive element into a series LC circuit, and a capacitive element into a shunt LC circuit, as show in Figure 4.6.

$$L_i = \left(\frac{\Omega_c}{\Delta\omega_0}\right)\gamma_0 g_i$$

$$C_i = \left(\frac{\Delta}{\omega_0\Omega_c}\right)\frac{1}{\gamma_0 g_i} \qquad \text{if } g_i \text{ represents the inductance} \qquad (4.21)$$

$$L_i = \left(\frac{\Delta}{\omega_0\Omega_c}\right)\frac{\gamma_0}{g_i} \qquad \text{if } g_i \text{ is the capacitance}$$

$$C_i = \left(\frac{\Omega_c}{\omega_0 \Delta}\right)\frac{g_i}{\gamma_0}$$

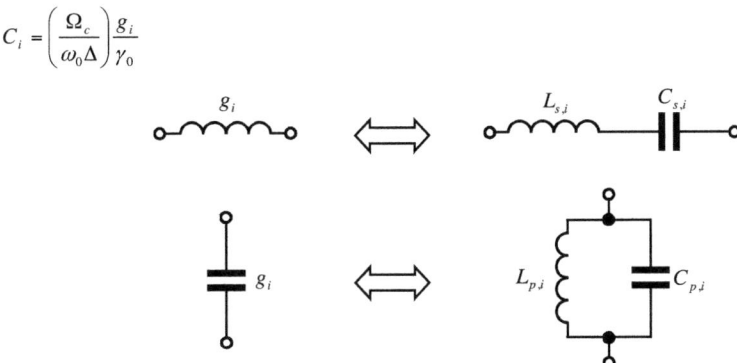

Figure 4. 6 Transformation of lowpass prototype filter elements to practical bandpass filter.

4.2.6. Immittance Inverters

There are two types of immittance inverters: impedance inverters (K–inverters) and admittance inverters (J–inverters). An ideal impedance inverter is characterized by its real characteristic impedance: K. If it is terminated by Z_L, then the input impedance at all frequencies is $Z_{in} = K^2/Z_L$. For the admittance inverter with the characteristic admittance J, termination Y_L, at the input we have $Y_{in} = J^2/Y_L$ (see Figure 4.7).

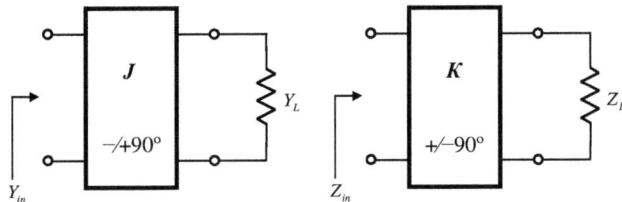

Figure 4. 7 Admittance and impedance inverters.

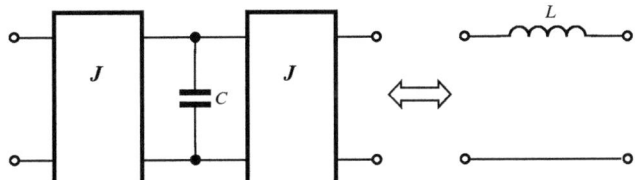

Figure 4. 8 Admittance inverters used to implemented a series inductance with a shunt capacitance.

Immittance inverters play an important role in the microwave filter design. They enable us to convert a structure, which is difficult to realize with microwave structures, into structure which is convenient for realization. Figure 4.8 depicts a simple example of transforming a two–port network with a series inductance into an electrically identical network with a shunt capacitance and admittance inverters. Immittance inverters exhibit a 90° phase shift of S_{21} parameter. For example the admittance inverter in Figure 4.9a exerts −90° phase shift, and the one in Figure 4.9b +90° phase shift. That is why we have the ±90° symbol in Figure 4.7.

Immittance inverters can be realized in a number of ways. The simplest inverter is a quarter–wavelength transmission line. Its characteristic impedance Z_c equals transmission line characteristic impedance $K = Z_c$, or $J = Y_c = 1/Z_c$. The transmission line is a narrow band immittance inverter around the frequency for which the transmission line has the quarter–wavelength.

Different versions of immittance inverters realized from lumped elements are shown in Figure 4.9. These structures are useful for converting the filter lowpass prototype, because the negative lumped elements ($-L_{i,i+1}$ and $-C_{i,i+1}$) can be absorbed in the shunt elements.

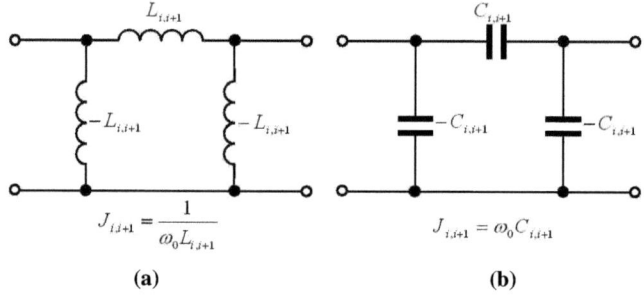

Figure 4. 9 Typical admittance inverters from lumped elements.

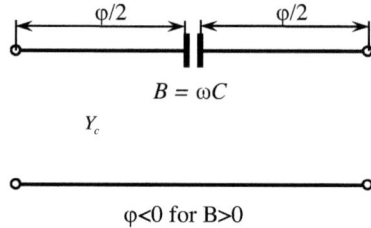

Figure 4. 10 Admittance inverter from lumped and transmission line elements.

Figure 4.10 depicts a practical immittance inverter, which consists of both a lumped element, and transmission lines. Let us note here that the inverter may have a transmission line with negative electrical length. The negative length will be subtracted from the adjacent line in the filter layout. The inverter characteristic admittance J, electrical length φ, and the capacitor susceptance B can be calculated using the following equations:

$$J = Y_0 \tan\left|\frac{\varphi}{2}\right| \tag{4.22}$$

$$\varphi = -\tan^{-1}\frac{2B}{Y_0}$$

$$\left|\frac{B}{Y_0}\right| = \frac{\dfrac{J}{Y_0}}{1-\left(\dfrac{J}{Y_0}\right)^2}$$

4.2.7. Filters with Immittance Inverters

The lowpass filter prototypes, shown in Figure 4.4, can be converted into the structure with admittance inverters shown in Figure 4.11. The new elements Y_0, Y_{n+1} and C_{a1} can be chosen arbitrarily, but the inverter characteristic admittance J_i has to be calculated according to the formulae in (4.23). In that case, the filter response will not change, compared to the prototype. The formulae in (4.23) can be obtained by equating the corresponding terms in the expanded input admittance of the two corresponding filter networks.

$$J_{0,1} = \sqrt{\frac{Y_0 C_{a1}}{g_0 g_1}}$$

$$J_{i,i+1} = \sqrt{\frac{C_{ai} C_{a(i+1)}}{g_i g_{i+1}}} \quad i = 1 \text{ to } n\text{-}1 \tag{4.23}$$

$$J_{n,n+1} = \sqrt{\frac{C_{an} Y_{n+1}}{g_n g_{n+1}}}$$

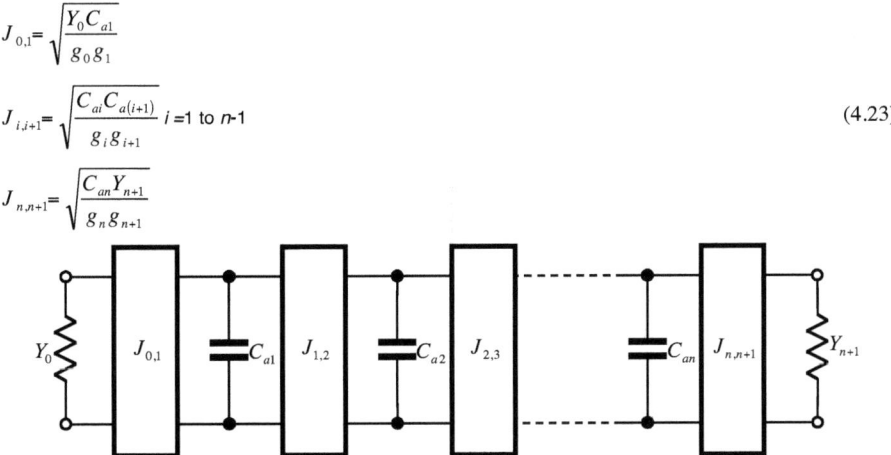

Figure 4.11 Lowpass filter as cascade of J–inverters and shunt capacitors.

The admittance inverters in Figure 4.11 are ideal, i.e. frequency invariable. This means that we can apply the bandpass transformation on the lowpass filter transformation in Figure 4.11. The shunt capacitor C_{ai} is transformed into the shunt parallel resonator, and the elements C_{pi} and L_{pi} are given by (4.24).

$$C_{pi} = \frac{\Omega_c}{\Delta\omega_0} C_{ai} \quad i = 1 \text{ to } n$$

$$L_{pi} = \frac{1}{\omega_0^2 C_{pi}} \quad i = 1 \text{ to } n \tag{4.24}$$

$$J_{0,1} = \sqrt{\frac{Y_0 \Delta \omega_0 C_{p1}}{\Omega_c g_0 g_1}}$$

$$J_{i,i+1} = \frac{\Delta \omega_0}{\Omega_c} \sqrt{\frac{C_{pi} C_{p(i+1)}}{g_i g_{i+1}}} \quad i = 1 \text{ to } n\text{-}1 \tag{4.25}$$

$$J_{n,n+1} = \sqrt{\frac{\Delta \omega_0 C_{pn} Y_{n+1}}{\Omega_c g_n g_{n+1}}}$$

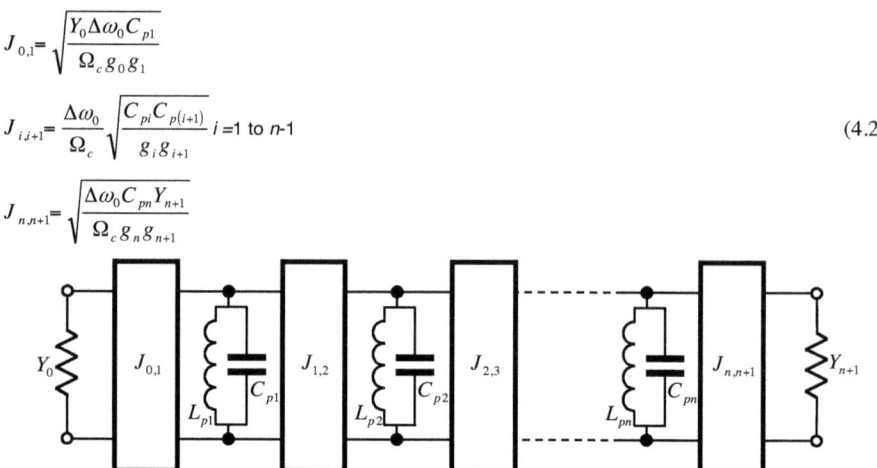

Figure 4.12 Bandpass filter as cascade of J–inverters and parallel LC resonators.

The inverter characteristic admittance has to stay unchanged by the bandpass transformation, and the formulae from (4.23) become (4.25) by substituting C_{ai} with $\Delta\omega_0 C_{pi}/\Omega_c$.

Up to this point we have been dealing with lumped elements. In order to realize microwave filters we have to use distributed elements such as microstrip or any other resonant structure. That is why we have to generalize the filter from Figure 4.12 by replacing the shunt parallel LC resonators with the susceptances of a distributed element [4.2]. The generalized bandpass filter with distributed elements is shown in Figure 4.13.

In the ideal case, the susceptance of the distributed circuit ($B(\omega)$) would equal the one of the parallel LC resonator in the whole frequency range. This is in reality not possible, so this is just an approximation around the bandpass filter centre angular frequency ω_0. The same is true for the inverters – the real ones will approximate ideal inverters only around the bandpass centre. In order to replace the parallel LC resonator with the distributed one, the distributed resonator susceptance and susceptance slope have to be equal to the corresponding values of the lumped resonator at ω_0. The susceptance slope is given by:

$$b = \frac{\omega_0}{2} \frac{dB(\omega)}{d\omega}\bigg|_{\omega=\omega_0} \tag{4.26}$$

The susceptance slope of the parallel LC resonator at the passband centre equals $\omega_0 C$. By replacing $\omega_0 C_{pi}$ in the equations in (4.25) with b_i we get the formulae (4.27) for the characteristic admittance of the inverters in Figure 4.13.

$$J_{0,1} = \sqrt{\frac{Y_0 \Delta b_1}{\Omega_c g_0 g_1}} \tag{4.27}$$

$$J_{i,i+1} = \frac{\Delta}{\Omega_c} \sqrt{\frac{b_i b_{i+1}}{g_i g_{i+1}}} \quad i = 1 \text{ to } n\text{-}1$$

$$J_{n,n+1} = \sqrt{\frac{\Delta b_n Y_{n+1}}{\Omega_c g_n g_{n+1}}}$$

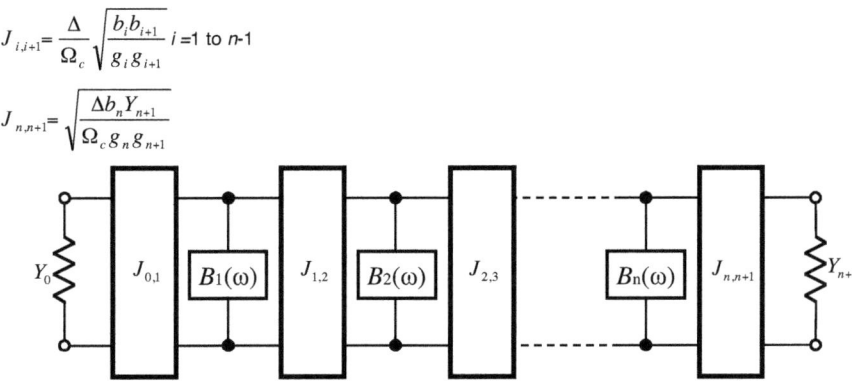

Figure 4.13 Bandpass filter as cascade of J–inverters and distributed elements.

4.2.8. Richards' Transformation

The filter prototypes presented in figures 4.4, 4.11 and 4.12, rely on lumped elements as resonators. When designing filters for the microwave range we have to rely on distributed elements such as transmission lines. In order to use developed filter prototypes, equivalence between lumped and distributed elements has to be established. This equivalence was established by Richards [4.3] and is called Richards' transformation. He showed that distributed networks which consist of transmission lines of equal electrical length and lumped resistors can be treated as lumped element *LCR* networks under transformation:

$$t = \tanh\left(\frac{lp}{v_p}\right) \tag{4.28}$$

where $p = \sigma + j\omega$ is the usual complex frequency variable, l is the transmission line element length, and v_p the transmission line wave phase velocity. For lossless networks $p = j\omega$ and the Richards' variable t becomes $t = j\tan\theta = j\Omega$, where $\theta = \omega l/v_p$ represents the electrical length.

The phase velocity for TEM transmission lines is independent of frequency, which makes the electrical length than proportional to the frequency, i.e. $\theta = \theta_0 \omega/\omega_0$, where θ_0 is the electrical length at the reference frequency ω_0. If we choose ω_0 so that $\theta_0 = \pi/2$, than the frequency mapping of ω into Ω performed by the Richards' transformation is given by:

$$\Omega = \tan\left(\frac{\pi}{2}\frac{\omega}{\omega_0}\right) \tag{4.29}$$

Figure 4.14 shows that frequency mapping from ω into Ω is periodic. The result of this is that a lowpass filter response will become periodic, with pass and stopbands periodically alternating with the period of $2\omega_0$ (shown in Figure 4.14.b). It is interesting to note that the filter may be seen as bandpass around $2\omega_0$, or bandstop around ω_0.

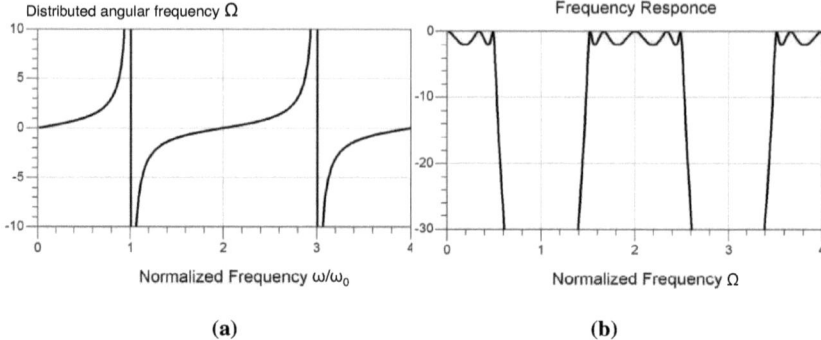

Figure 4.14 Richards' transformation of the real angular frequency ω into distributed angular frequency Ω (a), Chebyshev response after Richards' transformation (b).

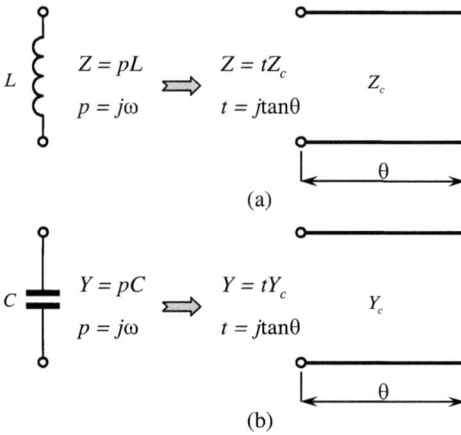

Figure 4.15 Richards' transformation of inductor (a) and capacitor (b) from p–plane into short– and open–circuited transmission line in the t–plane.

We have seen the effect of Richards' transformation on the Chebyshev response. This would happen if we replaced the lumped L and C elements with corresponding distributed elements in the t–plane. An inductor L with an impedance $Z = pL$ corresponds to short–circuited transmission line with an impedance $Z = tZ_c = jZ_c\tan\theta$, where Z_c is the characteristic impedance of the line. Likewise, a capacitor C with an admittance $Y = pC$ corresponds to open–circuited transmission line with an admittance $Y = tY_c = jY_c\tan\theta$, where Y_c is the characteristic admittance of the line. Fig 4.15 depicts this correspondence.

4.2.9. End–Coupled Microstrip Filters

The layout of an end–coupled microstrip bandpass filter is shown in Figure 4.16. It consists of n open–end microstrip resonators. The resonators have length of approximately one half of guided

Chapter 4   Image–rejection Filter

wavelength at the centre angular frequency ω_0. The coupling between two resonators is through the gap between their adjacent open–ends. The coupling is capacitive and can be seen as a J–inverter, as in Figure 4.10. These J–inverters reflect high impedance to the ends of both resonators, causing the resonators to exhibit a shunt–type resonance [4.2]. This means that the filter in Figure 4.12. can be used as an equivalent circuit for end–coupled filters around ω_0. Using formulae (4.22), (4.25) and (4.26) we get the general design equations for the end–coupled filter [4.4]:

$$\frac{J_{0,1}}{Y_0} = \sqrt{\frac{\pi}{2} \frac{\Delta}{g_0 g_1}}$$

$$\frac{J_{i,i+1}}{Y_0} = \frac{\pi \Delta}{2} \frac{1}{\sqrt{g_i g_{i+1}}} \quad i = 1 \text{ to } n\text{-}1 \tag{4.30}$$

$$\frac{J_{n,n+1}}{Y_0} = \sqrt{\frac{\pi}{2} \frac{\Delta}{g_n g_{n+1}}}$$

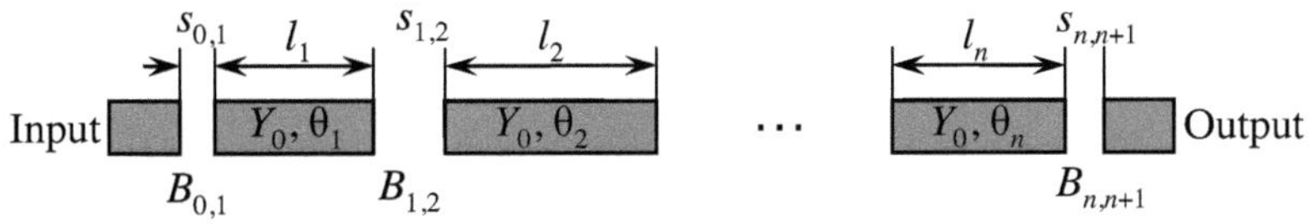

Figure 4.16 Layout of an end–coupled microstrip bandpass filter.

Before continuing with the end–coupled filter design procedure, let us first analyze two types of microstrip discontinuities: the open–end (Figure 4.17a) and the gap (Figure 4.17b). They are of interest for the design procedure and there equivalent circuits will be examined.

The electromagnetic field at the microstrip open–end extends further due to the effect of the fringing field. This effect is modeled either with a shunt capacitor C_p or with an equivalent length of the transmission line Δl (Figure 4.17a). The relation between these two parameters is given by [4.5]:

$$\Delta l = \frac{c C_p}{Y_c \sqrt{\varepsilon_{re}}} \tag{4.31}$$

where c is the light velocity in free space, ε_{re} the effective dielectric constant. The closed–form expression for Δl is given in [4.6].

The equivalent circuit for the microstrip gap is shown in Figure 4.17b. The calculation of capacitances C_p and C_g are given in [4.7]. The other way to calculate these capacitances is by means of EM simulation. If we assume the Y–parameters as the result of the EM simulation, than C_p and C_g are given by:

$$C_g = -\frac{\text{Im}(Y_{21})}{\omega_0} \tag{4.32}$$

$$C_p = \frac{\operatorname{Im}(Y_{11} + Y_{21})}{\omega_0}$$

where Im(x) represents the imaginary part of x.

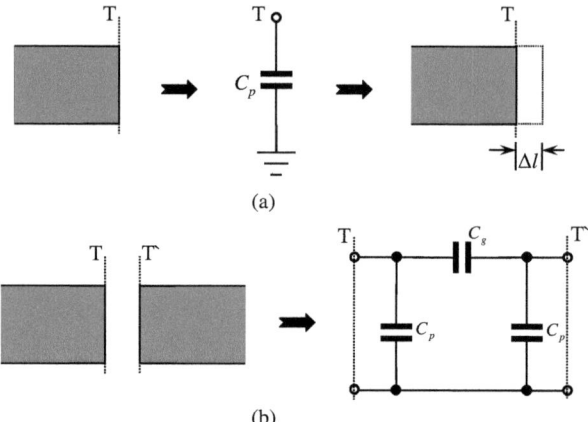

Figure 4.17 Microstrip discontinuities and their equivalent circuits. Open–end (a) and gap (b).

Now we can present the design flow for the end–coupled filters.

1. Filter response type and order is chosen based on the filter specifications.

2. Prototype filter elements g_i are calculated as given in section 4.2.4.

3. The normalized characteristic admittance is calculated as given in (4.30).

4. The electrical length of the resonator i, and the gap series susceptance $B_{i,i+1}$ can be calculated based on (4.22) as:

$$\left|\frac{B_{i,i+1}}{Y_0}\right| = \frac{\frac{J_{i,i+1}}{Y_0}}{1 - \left(\frac{J_{i,i+1}}{Y_0}\right)^2} \tag{4.33}$$

$$\theta_i = \pi - \frac{1}{2}\left[\tan^{-1}\left(\frac{2B_{i-1,i}}{Y_0}\right) + \tan^{-1}\left(\frac{2B_{i,i+1}}{Y_0}\right)\right] \text{ [rad]}$$

As we can see in the second expression in (4.33), the resonator electrical length absorbs the negative lengths of the J–invertors.

The series gap capacitance is calculated from:

$$C_g^{i,j+1} = \frac{B_{i,j+1}}{\omega_0} \qquad (4.34)$$

5. In order to determine the gap dimension $s_{i,i+1}$, EM simulation has to be implemented. The simulation is performed to find the right gap dimension that gives the same value for $C_g^{i,j+1}$ (Equation (4.32)) as the one calculated from (4.33). EM simulation results are than used to obtain $C_p^{i,j+1}$ from (4.32).

6. The physical length of resonators is given by:

$$l_i = \frac{\lambda_{g0}}{2\pi}\theta_i - \Delta l_i^{e1} - \Delta l_i^{e2}$$

$$\Delta l_i^{e1} = \frac{\omega_0 C_p^{i-1,i}}{Y_0}\frac{\lambda_{g0}}{2\pi} \qquad (4.35)$$

$$\Delta l_i^{e2} = \frac{\omega_0 C_p^{i,i+1}}{Y_0}\frac{\lambda_{g0}}{2\pi}$$

where Δl_i^{e1} and Δl_i^{e2} represent the effective lengths of $C_p^{i-1,i}$ and $C_p^{i,i+1}$ shunt capacitances at the ends of resonator i, and λ_{g0} is the guided wavelength at ω_0.

4.2.10. Parallel–Coupled Microstrip Filters

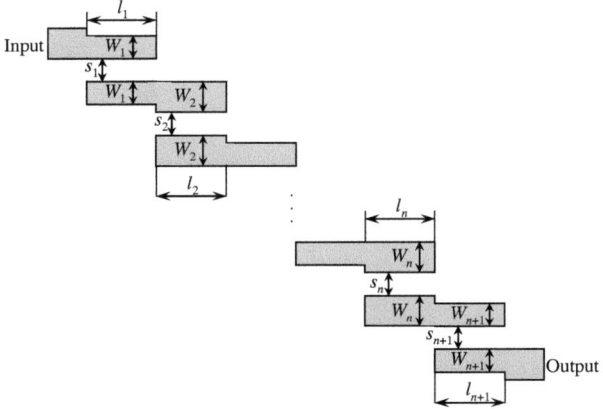

Figure 4. 18 Layout of a parallel–coupled microstrip bandpass filter.

The layout of a parallel–coupled (or edge–coupled) microstrip bandpass filter is shown in Figure 4.18. It consists of n open–end microstrip resonators, with length of approximately one half of guided wavelength at the centre angular frequency ω_0. The resonators are positions in such a way that the adjacent resonators are parallel along half of their length, as shown in Figure 4.18. The coupling between two resonators is much stronger compared to the coupling through the gap in the end–coupled filters. This makes the parallel–coupled filters better solution for the broadband filters. The parallel–coupled filters are also more compact than the end–coupled filters.

The coupling structure is the coupled transmission line. The behavior of two coupled lines is characterized by two characteristic impedances. The even–mode impedance Z_{0e} is defined by exiting both lines with the same signal amplitude and phase. The odd–mode impedance Z_{0o} is defined by exiting both lines with signals of equal amplitude and phase difference of 180°. The maximum coupling between the lines is achieved for the electrical length of the coupled section of 90°.

The formulae for the calculation of the even–mode impedance Z_{0e} and the odd–mode impedance Z_{0o} of coupled microstrip lines are given in [4.8], and more accurate formulae in [4.9]. These formulae are rather complex and will not be reprinted here. The characteristic impedances Z_{0e} and Z_{0o} can be calculated using some of the commercially available software tools, such as Agilent's ADS LineCalc. This software allows both the analysis of a certain structure on a predefined substrate and its synthesis for the required parameters.

The design equations for the parallel–coupled filter are given in (4.30) [4.4]. They are the same as for the end–coupled filter because both filters have the same lowpass network representations.

Now we can present the design flow for the parallel–coupled filters.

1. Filter response type and order is chosen based on the filter specifications.

2. Prototype filter elements g_i are calculated as given in section 4.2.4.

3. The normalized characteristic admittance is calculated as given in (4.30).

4. The characteristic impedances Z_{0e} and Z_{0o} of the coupled i and $i+1$ microstrip line are given by [4.4]:

$$(Z_{0e})_{i,i+1} = \frac{1}{Y_0}\left[1 + \frac{J_{i,i+1}}{Y_0} + \left(\frac{J_{i,i+1}}{Y_0}\right)^2\right] \quad i = 0 \text{ to } n$$

$$(Z_{0o})_{i,i+1} = \frac{1}{Y_0}\left[1 - \frac{J_{i,i+1}}{Y_0} + \left(\frac{J_{i,i+1}}{Y_0}\right)^2\right] \quad i = 0 \text{ to } n \tag{4.36}$$

5. The dimensions of the microstrip lines are calculated using ADS LineCalc tool, for the given substrate definition and calculated characteristic impedances.

4.2.11. Hairpin Microstrip Filters

The layout of a hairpin (or hairpin–line) bandpass microstrip filter is show in Figure 4.19. The main advantage of the hairpin filter compared to the end–coupled and parallel–coupled ones is that it is very compact. It can be formed by bending the resonator of the parallel–coupled filter into the

"U" or hairpin shape. Consequently, the design equations for the parallel–coupled filter in (4.30) can be used for the hairpin filter. The hairpin resonator will, however, have additional coupling between the two lines of the same resonator, affecting its properties and the filter response. Hence, a different, more accurate design procedure is applied.

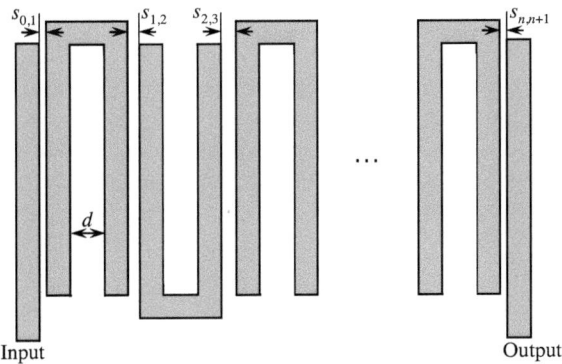

Figure 4. 19 Layout of a hairpin microstrip bandpass filter.

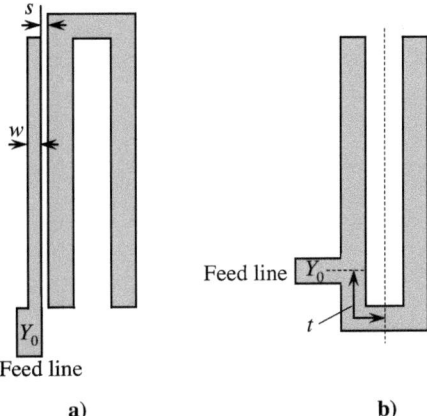

Figure 4. 20 Typical input or output couplings of a coupled resonator filter. Coupled–line coupling a), tapped–line coupling b).

The theory for the hairpin filters is a general theory of coupled resonator filters, which relies on determining the coupling coefficients between the resonators M and external quality factors of the resonators at the input and the output, Q_e. The design equations for the hairpin filter are given by [4.4]:

$$Q_{e1} = \frac{g_0 g_1}{\Delta} \qquad (4.37)$$

$$M_{i,i+1} = \frac{\Delta}{\sqrt{g_i g_{i+1}}} \quad i = 1 \text{ to } n-1$$

$$Q_{en} = \frac{g_n g_{n+1}}{\Delta}$$

where Q_{e1} and Q_{en} are the input and output external quality factor, respectively, and $M_{i,i+1}$ the coupling coefficient between the ith and $i+1$ resonator.

There are two typical ways of coupling the feed line to the first or last resonator: parallel–coupling of the feed line and the resonator (4.20a) and feed line direct tapping of the resonator.(4.20b). The external quality factor can be obtained using the EM simulation [4.4], and analyzing the phase graph of the S_{11} parameter. The Q_e value is given by:

$$Q_e = \frac{\omega_0}{\omega_{-90°} - \omega_{+90°}} \tag{4.38}$$

where $\omega_{-90°}$, ω_0 and $\omega_{+90°}$ are frequencies at which the is the frequency for which the S_{11} parameter phase equals –90, 0 and +90 degrees (see Figure 4.21).

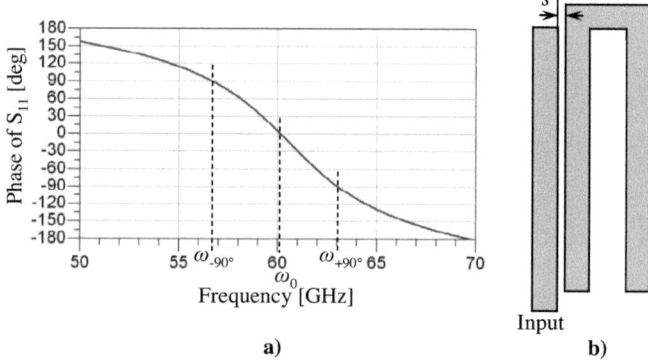

Figure 4.21 The external quality factor parameter extraction.

The extraction of the coupling coefficient parameter M also relies on EM simulation. Two resonators, shown in Figure 4.22b, are simulated for different values of separation s. Input and output feed lines can be, for example, parallel–coupled. In that case the feed line – resonator separation must be high to minimize the feed line effect on the resonant frequencies of the coupled resonators. The other option is to connect ports directly to the resonators, but the port impedance has to be made very high to minimize the effect on the resonant frequencies. Typical frequency response is presented in Figure 4.22a. The coupling coefficient can be calculated by:

$$M = \frac{f_2^2 - f_1^2}{f_2^2 + f_1^2} \tag{4.39}$$

where f_1 and f_2 represent the lower and higher resonant frequency, respectively.

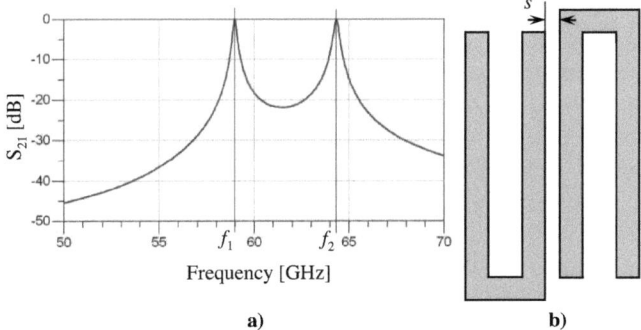

Figure 4. 22 The coupling coefficient parameter extraction.

Now we can present the design flow for the parallel–coupled filters.

1. Filter response type and order is chosen based on the filter specifications.

2. Prototype filter elements g_i are calculated as given in section 4.2.4.

3. The coupling coefficients M and external quality factors are calculated as given in (4.37).

4. Feed line and resonator are simulated, and external quality factor is calculated as given in (4.38). Based on these simulations, the appropriate separation s is chosen for the required values of Q_{e1} and Q_{en}.

5. Two resonators are simulated, and coupling coefficient is calculated as given in (4.39). Based on these simulations, the appropriate separation s is chosen for the required values of $M_{i,i+1}$.

4.2.12. Selective Microstrip Filters with Transmission Zeros

Some applications, especially some wireless communications standards, require very selective filters. Some advanced filter structures, such as quadruplet and trisection filters, offer improved selectivity by introducing transmission zeros close to the passband. This kind of filters are a good alternative to standard Chebyshev filters of high order, or elliptic filters, which are more difficult for realization. The transfer function of a filter with a single pair of transmission zeros is given by:

$$|S_{21}(\Omega)| = \frac{1}{\sqrt{1+\varepsilon^2 F_n^2(\Omega)}}$$

$$F_n(\Omega) = \begin{cases} \cos\left((n-2)\cos^{-1}\Omega + \cos^{-1}\left(\frac{\Omega_a\Omega-1}{\Omega_a-\Omega}\right) + \cos^{-1}\left(\frac{\Omega_a\Omega+1}{\Omega_a+\Omega}\right)\right), & |\Omega| \leq 1 \\ \cosh\left((n-2)\cosh^{-1}\Omega + \cosh^{-1}\left(\frac{\Omega_a\Omega-1}{\Omega_a-\Omega}\right) + \cosh^{-1}\left(\frac{\Omega_a\Omega+1}{\Omega_a+\Omega}\right)\right), & |\Omega| > 1 \end{cases} \quad (4.40)$$

where Ω is angular frequency normalized to the passband cutoff frequency of the lowpass prototype filter, ε is the ripple constant ($\varepsilon = \sqrt{10^{L_R/10} - 1}$), L_R the return loss, and $\pm\Omega_a$ represents the location of the transmission zeros, i.e. attenuation poles. We can note that if $\Omega_a \rightarrow \infty$ the transfer function becomes standard Chebyshev function. Figure 4.23 gives a comparison between a 4th order Chebyshev filter and filters with transmission zeros at $\Omega_a = 1,4, 1,8$. Improved skirt selectivity is evident. However, out–of–band suppression is deteriorated.

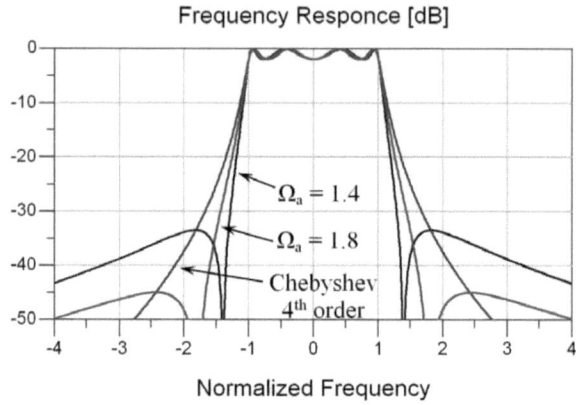

Figure 4. 23 Frequency response of a lowpass 4th order Chebyshev filter (red), with transmission zeros at 1.4 (blue) and 1.8 (pink) of the normalized frequency.

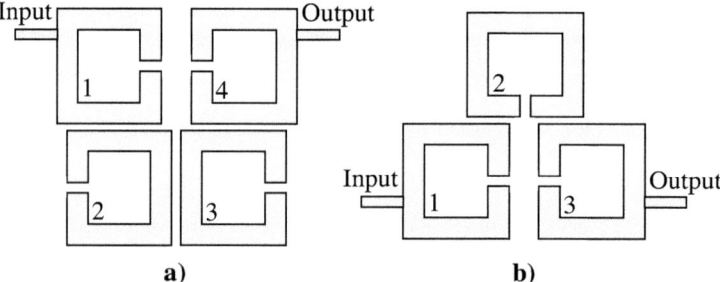

Figure 4. 24 An example of a four–pole microstrip quadruplet filter (a), and a three–pole microstrip trisection filter (b).

Figure 4.24 depicts two bandpass, microstrip filters with transmission zeros: quadruplet filter (a) and trisection filter (b). From the quadruplet filter layout we can see the principal for the realization of transmission zeros. Resonator pairs 1 and 2, 2 and 3 and 3 and 4 are strongly coupled. Resonators 1 and 4 are weakly coupled. The signal which passes through resonators 1,2, 3 and 4 is subtracted from the signal which goes directly from resonator 1 to 4. At frequencies where these signals fully subtract we have zero transmission, i.e. these frequencies are transmission zeros. The quadruplet structure in Figure 4.24a gives two transmission zeros around the passband, and the trisection structure in 4.24b gives one transmission zero.

Design procedure is given in [4.4]. The first part of the procedure is the calculation of the external quality factors and coupling coefficients. This calculation depends on the type of the filter. The second part of the procedure is to calculate resonator separation for a given coupling factor and feed–line coupling for a given external quality factor. This part is common to all types of filters, and it was already explained in the previous section on hairpin filters.

4.3. Filter Losses

The filters analyzed so far were considered lossless. In reality all practical filters consist of lossy, lumped or distributed elements. A lossy inductor is conventionally presented as a series of an ideal inductor L and resistor R (Figure 4.25a). A lossy capacitor equivalent circuit is shown in Figure 4.25b. The inductor and capacitor unloaded quality factors Q_u at the angular frequency ω is given by:

$$Q_u = \frac{\omega L}{R} \quad (4.41)$$

$$Q_u = \omega C R \quad (4.42)$$

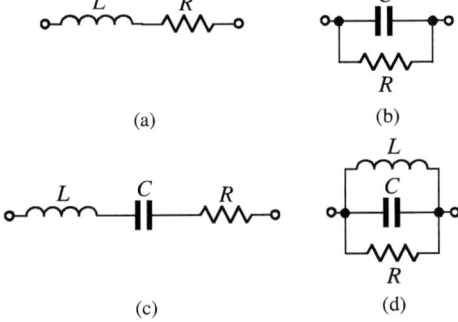

Figure 4.25 Circuit representations of lossy reactive elements (a) and (b) and series and parallel resonators (c) and (d).

For a lowpass and highpass filter ω is usually the cutoff frequency, and for a bandpass and bandstop filter it is the centre frequency. Lossy series and parallel resonator equivalent circuits are presented in Figure 4.25c and d, respectively. Their quality factors can be calculated with (4.41) and (4.42), for the series and parallel resonator, respectively. The angular frequency ω is in this case the resonant frequency, i.e. $\omega = 1/\sqrt{LC}$.

4.3.1. Effects of Filter Losses on the Lowpass Frequency Response

It is of interest to examine the effect of the dissipation on the filter response. For that purpose let us consider a simplified case where all elements have the same unloaded quality factor. In that case, after designing an ideal filter, the resistive elements in Fig 4.25 can be calculated from (4.41) and (4.42). Figure 4.24 shows filter response of a lowpass 5th order Butterworth filter for $Q_u = 10, 50, 100$ and $Q_u \to \infty$.

We can note two effects of the dissipation. One is the increase of the insertion loss by a constant amount starting from dc. The second effect is the reduction of selectivity through the rounding of the response curve around the cutoff frequency, and reduction of the passband. We may note that the effects are relatively significant only for $Q_u < 50$.

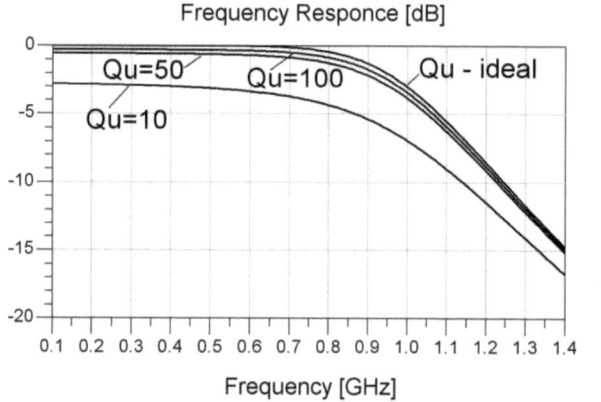

Figure 4. 26 Frequency response of a lowpass 5th order Butterworth filter with lumped elements for different quality factor values of the filter elements.

The formula which estimates the insertion loss at zero frequency for a ladder–type lowpass filter is given by [4.2]:

$$\Delta L_{A0} = 4.434 \sum_{i=1}^{n} \frac{\Omega_c}{Q_{ui}} g_i \; [\text{dB}] \tag{4.43}$$

where Ω_c is the normalized filter cutoff frequency, g_i is the element value with the unloaded quality factor Q_{ui} given at the angular cutoff frequency ω_c.

4.3.2. Effects of Filter Losses on the Bandpass Frequency Response

Let us now examine the effects of dissipation on a bandpass filter. The formula in (4.31) can be modified for the bandpass filter, and is given by [4.2]:

$$\Delta L'_{A0} = 4{,}434 \sum_{i=1}^{n} \frac{\Omega_c}{\Delta Q_{ui}} g_i \; [\text{dB}] \qquad (4.44)$$

where $\Delta L'_{A0}$ represents the increase in insertion loss at the centre frequency, and Δ the fractional bandwidth – as defined in (4.20). We may note here that the increase in insertion loss for a bandpass filter is even more significant compared to the lowpass filter, because for the same filter prototype elements it is larger by the factor $1/\Delta$. For a narrowband filter this is a large factor.

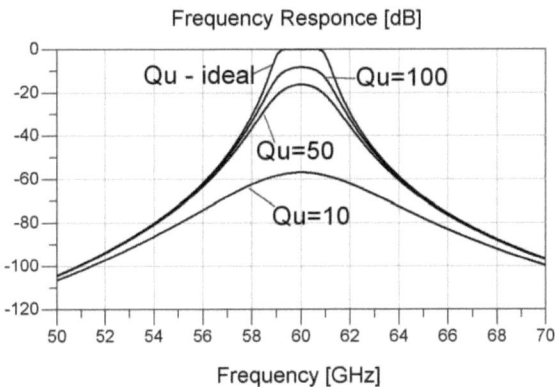

Figure 4. 27 Frequency response of a 60 GHz bandpass filter with lumped elements for different quality factor values of the filter elements.

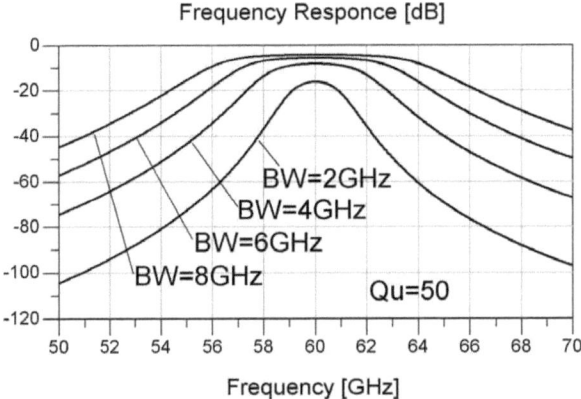

Figure 4. 28 Frequency response of a 60 GHz bandpass filter with lumped elements for different bandwidth values.

Figure 4.27 shows the frequency response a bandpass 5th order Butterworth filter with lumped elements, centre frequency is 60 GHz, and bandwidth 2 GHz. As for the lowpass filter case, all resonators are assumed to have the same unloaded quality factor value $Q_u = 10, 50, 100$ and $Q_u \rightarrow \infty$. Even for $Q_u = 100$ the insertion loss at ω_0 is as high as 8.3 dB, and for $Q_u = 10$ it is 57 dB. Strong deterioration of the selectivity is also obvious at Figure 4.27.

Increasing the filter bandwidth will reduce the insertion loss, as can be seen in Figure 4.28. This will, however, strongly reduce the selectivity.

4.3.3. Microstrip Filter Losses

The microstrip filters, which are of interest for the 60 GHz range, exhibit three kinds of losses: the conductor loss, the dielectric loss and the radiation loss. The signal propagation on a lossy transmission line is characterized by the attenuation constant α, which is given in dB per unit of length. The attenuation constant is a sum of attenuation constants referring to each loss source, i.e. $\alpha = \alpha_c + \alpha_d + \alpha_r$.

The conductor loss of a transmission line can be estimated by [4.10]:

$$\alpha_c = \frac{8{,}686 R_s}{Z_c W} \tag{4.45}$$

where Z_c is the microstrip characteristic impedance, W the width of the microstrip line and R_s is the surface resistance of the signal line and the ground plane. The surface resistance is given by:

$$R_s = \sqrt{\frac{\omega \mu_0}{2\sigma}} \tag{4.46}$$

where μ_0 is permeability of free space, σ the conductor material conductivity and ω the angular frequency. This loss is usually not significant for filters realized on PCBs, because the metal lines have large enough cross–section reducing the signal attenuation per unit of length, but for an integrated filter the dimensions are much smaller and this loss becomes more important. The skin effect and surface roughness may significantly increase the conductor loss for high frequency signals. The skin effect causes the current to flow more on the conductor surface, and less through the centre, effectively reducing the conductor cross–section and increasing the loss. The second effect is the consequence of the fabrication process where the conductor surface is not flat, but with bumps and holes. This increases the path of the high frequency currents, effectively increasing the conductor surface resistance.

The dielectric loss is given by:

$$\alpha_d = 8{,}686\pi \left(\frac{\varepsilon_{re} - 1}{\varepsilon_r - 1} \right) \frac{\varepsilon_r}{\varepsilon_{re}} \frac{\tan \delta}{\lambda_g} \text{ [dB/unit length]} \tag{4.47}$$

where tanδ is the loss tangent, ε_r the relative dielectric constant, ε_{re} the effective dielectric constant and λ_g is the guided wavelength.

The radiation loss of a transmission line is very small. It becomes significant only when the distance between the strip and ground is comparable to the wavelength. However, discontinuities, such as open ends can induce radiation. Radiation is stronger at higher frequencies. The microstrip filters radiate into the open air, and when they have a metallic enclosure the radiation is absorbed by it. That loss is called the housing loss.

4.4. Design of the 60 GHz Image–Rejection Filter

4.4.1. Specifications of the Image–rejection Filter

As discussed in the second chapter, the result of the mixing process of the IF input signal around 5.5 GHz and 56 GHz PLL signal is the wanted signal at 61.5 GHz and the image at 51.5 GHz. The purpose of the image–rejection filter is to suppress this signal. The calculation showed negligible increase of the BER for image suppression of 30 dB or more. The required image suppression of 30 dB does not have to be fully achieved by the filter because the other components are also more or less selective. The preamplifier was designed earlier and it is broadband. The mixer is also broadband, but the image is somewhat stronger than the signal. It can be assumed that the combination of the mixer and preamplifier has the approximately same conversion gain for both the signal and the image. The PA was not measured before the filter was designed, but it was designed as selective, with 10 dB selectivity. This means that the minimum filter selectivity is 20 dB. Target selectivity of 25 dB is chosen to compensate for the possible lower selectivity of the measured filter.

Filter insertion loss is also important and should be as low as possible, but we should keep in mind that it can be compensated by more amplification from the preamplifier and the PA. Bandwidth of the modulated signal is 400 MHz. The required bandwidth of the filter should be 2 GHz to compensate for the possible frequency shifts of the real filter.

There are three more very important aspects of the filter. One is the area the filter occupies. The area is very important because this is an integrated filter and chip area is expensive. The second aspect concerns the shape of the filter. A square shape is preferred, because it is more likely to fit in the chip layout, compared with a narrow but long filter layout. We should, however, note that this depends on the specific design. For our case square shape is preferred. The last aspect is the position of the input and the output. Output could, for example, be on the "wrong" side, and the signal would have to propagate on a transmission line parallel to the filter, introducing additional coupling and affecting filter response. The "correct" position of the input and output is dependant on the specific design. For our case it is preferred to have them at one side, or at the adjacent corners.

Two integrated filters for the IEEE 802.15.3c standard [4.11] were developed. The required pass band is from 57–to–66 GHz, and the image is at 60% of the signal frequency (IF is not fixed), i.e. the highest image frequency is 40 GHz.

4.4.2. The State–of–the–Art in 60 GHz Filters

There is a very broad activity in the design of filters for different standards and frequency ranges. However, there is very limited work done on the filters for 60 GHz applications. When the work on the integrated filter for the transmitter started, there was no integrated filter published. Table 4.I summarizes filters for 60 and 77 GHz.

Reference	Insertion Loss [dB]	3-dB Bandwidth [GHz]	Centre Frequency [GHz]	Image Rejection @offset	Return Loss [dB]	Technology	Size [mm²]
[4.12]	4	4.5	61	20dB @6GHz	13.6	LTCC	-
[4.13]	3.1	2	60	30dB @3GHz	17.2	PCB	4×2
[4.14]	3.4	8.5	58	20dB @11GHz	13.1	MEMS on GaAs	2.3×2.3
[4.15]	0.5	2.5	60.5	30dB @3.5GHz	11.2	NRD guide	15×2.5
[4.16]			77	20dB @6GHz			
[4.17]	6.4	12	77	28dB @20GHz	12	Integrated	0.11×0.06

Table 4. I Comparison of 60 and 77 GHz Filters.

Papers [4.12] and [4.13] present 60 GHz on-board filters with 3 to 4 dB insertion loss. [4.13] has very good selectivity of 30 dB@3GHz offset. Filters in [4.14] and [4.15] were produced in special technological procedures, which are not useful for our application. Only the paper [4.16] deals with an integrated filter.

4.4.3. Filter Simulation and Substrate Definition

Filters were designed for a 0.25 μm process with five aluminum metal layers. Top metal layer is the thickest (3 μm), and it is used for filter design to minimize ohmic losses. The bottom metal layer is used for ground. Substrate cross–section is presented in Figure 4.29a. There is a 1.5 μm thick passivation layer on top of the top metal layer. This passivation layer is not leveled as show in Figure 4.29a.

Electromagnetic (EM) simulations were done with Agilent's simulator Momentum. It is a 2.5 dimensional simulator. It is capable to simulate only planar structures, so the substrate definition regarding the passivation layer was used as shown in Figure 4.29b. This kind of a substrate is good for simulating TLs where most of the EM field is confined between the signal line and ground layer. For microstrip filters, most of the EM fields are between the resonators (in the top metal layer), and an accurate substrate definition is needed for this space. This space is partly filled with SiO_2 and partly with air. The shape of Sio2-air profile is not planar and it depends on the spacing between the resonators. That is why it is not possible to accurately describe the substrate for microstrip filter simulation. To solve this problem, a fitted substrate definition was created based on the measurement results of the first microstrip filter that was produced. This substrate definition is show in Figure 4.29c. This substrate has proven to be accurate enough for different versions of microstrip filters.

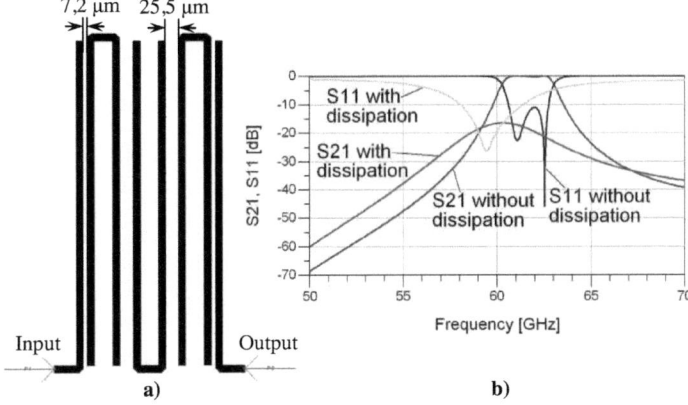

Figure 4. 29 Substrate cross–section (not to be scaled): a) physical, b) for Momentum simulations, c) fitted substrate for measured data.

4.4.4. Filter Design

There is a large number of microstrip topologies which could be used for realization of bandpass filters for 60 GHz range. Since the goal is filter integration, only compact topologies are of interest. Hairpin filter is in this sense a good candidate. Trisection filter also has an interesting topology. It exhibits one transmission zero resulting in an asymmetric attenuation pattern for the upper and lower stopband. This is of our interest, because only one signal has to be suppressed, and a design with a transmission zero at the image frequency would have very good selectivity.

Figure 4. 30 a) 3rd order Chebyshev filter layout. b) Simulated filter S parameters with and without dissipation.

Let us first examine the hairpin filter. The filter is designed for the specification given in section 4.4.1 and according to the procedure given in section 4.2.11. Chebyshev filter type is chosen because it has better selectivity compared to the Butterworth type. Elliptic filter has even better selectivity, but it can't be realized as a hairpin filter. We should note here that the formula for filter order (4.12) is not valid for a microstrip bandpass filter, but only for a ladder–type lumped element filter. Simulation shows that minimum required filter order is three. 3rd order Chebyshev prototype filter elements are: $g_0 = g_4 = 1$, $g_1 = g_3 = 1.0316$ and $g_2 = 1.1474$. This gives the input and output external quality factor parameters: $Q_{e1} = Q_{e3} = 31.5$ and coupling coefficient parameters: $M_{12} = M_{23} = 0.0301$.

The resonator length should be half of wavelength at 61.5 GHz. EM simulation shows that this corresponds to 1330 μm. The external quality factor parameter and coupling coefficient parameter are extracted as explained in section 4.2.11. and feed line–resonator and resonator–resonator separation are extrapolated from the results. The filter layout is shown in Figure 4.30a. The filter size is $0.28 \times 0.65 = 0.18$ mm². Simulation results without any kind of loss shows 3 dB bandwidth of approximately 2.7 GHz around 61.5 GHz, and image rejection of 63 dB (Figure 4.30b). The same graph shows filter response with all kinds of loss included. We see strong deterioration of the filter response characteristics. Insertion loss is 16.5 dB, bandwidth is 4.2 GHz centered around 60.4GHz, and image rejection is 42 dB. Filter center frequency was shifted by 1.8%. Reducing resonator length by the same percentage would shift the passband center frequency back to 61.5 GHz.

There are three reasons why the deterioration is so strong. First, integrated metal lines have very small cross–section making ohmic losses much larger compared to PCB board lines. Second, radiation loss around 60 GHz is greater than for lower frequencies. Third, narrow bandpass filters are much more susceptible to loss compared to lowpass filters (see Figure 4.26 and 4.27).

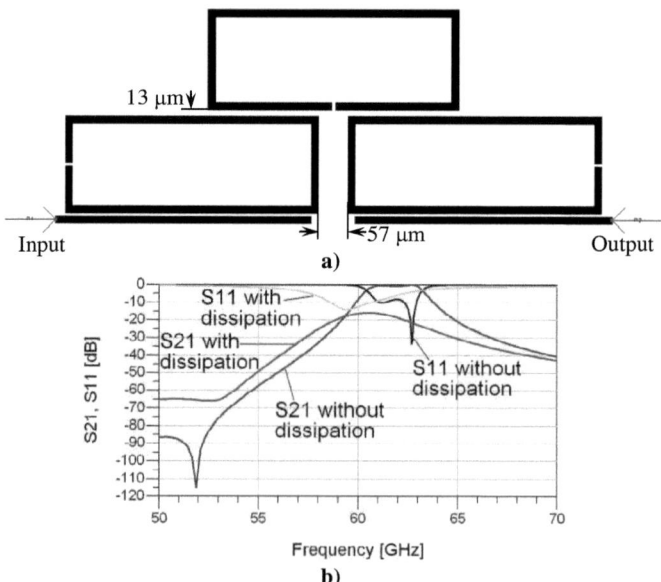

Figure 4. 31 a) Trisection filter layout. b) Simulated filter S parameters with and without dissipation.

This filter with 43 dB simulated image rejection is much more selective than necessary. This leaves us with an option to make the filter more broadband, which will improve insertion loss, but reduce image rejection (see Figure 4.28).

Let us now examine the filter with a trisection topology. As mentioned in the section 4.2.12, design procedure is given in [4.4], and filter was designed for the specifications given in section 4.4.1. Figure 4.31 shows a layout and simulated S parameters of a trisection filter. Filter size is 1.04×0.42 = 0.44 mm^2. Simulation results without dissipation show 3 dB bandwidth of approximately 2.8 GHz around 61.8 GHz, and image rejection of more than 90 dB (Figure 4.31b). The same graph shows filter response with dissipation. Insertion loss with dissipation is 16.5 dB, bandwidth is 3 GHz centered around 60.5GHz, and image rejection is 49 dB. Filter center frequency can be shifted to 61.5 GHz by reducing resonator length by 1.6%.

The Chebyshev and trisection filter gave very similar results, with trisection filter having better image rejection. Both have very high insertion loss, and would have to be redesigned as broadband for lower insertion loss. Broadband filters need strong coupling between resonators, i.e. resonators have to be next to each other in close proximity. Hairpin resonators are placed in parallel and have stronger coupling compared to trisection resonators, which are only partly placed in parallel. Therefore, hairpin filter are better suited for broadband design. Hairpin filter is also smaller and it would fit better in the transmitter layout due to the position of the input and output.

4.4.4.1. Broadband Hairpin Filter

As discussed already, the hairpin filter had to be redesigned as a broadband to reduce the insertion loss. The filter was designed so that the lowest transmission pole equals the signal frequency, i.e. 61.5 GHz (see Figure 4.32b). For larger filter bandwidth, the insertion loss decreases, but the filter is also less selective. The highest bandwidth that satisfies the condition of 25 dB image rejection is 15 GHz. EM Momentum simulations were done for the substrate shown in Figure 4.29b. Minimum value of the insertion loss achieved in simulation is 7.3 dB. This is a large improvement compared to the narrowband filter with 16.5 dB insertion loss. Filter layout photo is shown in Figure 4.32a, together with the deembeding structures: short, open and thru, below the filter. Feed lines are 50 Ohm transmission lines, 14 µm wide. The resonator lines are 30 µm wide. Wider resonator lines were chosen because they give better selectivity.

The filter was produced and measured. Measured data was shown in Figure 4.33. The position of the poles and image rejection match well with the simulation, but there is significant mismatch in the value of the insertion loss. The lower transmission pole is shifted up by 1 GHz, which can be easily compensated by increasing the resonator length proportionally to the frequency shift. That is way we should observe the image rejection and insertion loss also with 1 GHz shift. Image rejection is 23 dB (62.5 to 52.5 GHz). Insertion loss is 11 dB (at 62.5 GHz). Both in simulation and measurement the return loss is very low, just −3 dB. The filter dimensions are 0.6x0.4 mm^2.

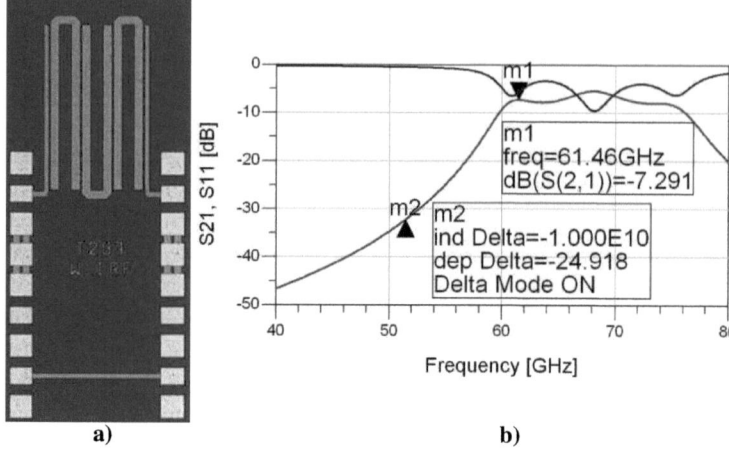

Figure 4.32 a) Photo of the manufactured filter with the deembeding structures: short, open and thru. b) Simulated filter S parameters.

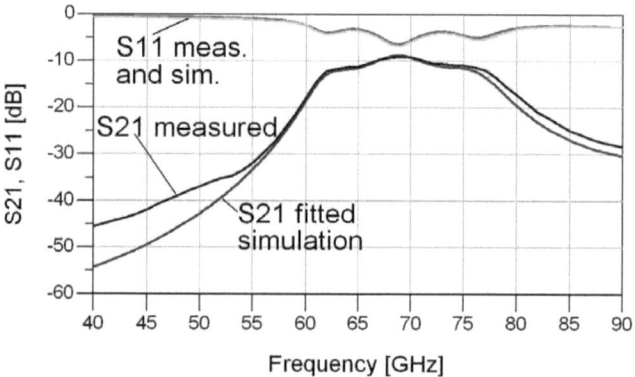

Figure 4.33 S parameters of the hairpin filter: measured and simulated with fitted substrate.

The substrate definition was changed to fit the measurement results (see Figure 4.29c). The new substrate definition parameters don't have physical values. They are just used to better predict measurement results. This substrate definition was used for the subsequent filter designs.

This filter was used for the TX version I.

4.4.4.2. Broadband Parallel–Coupled Asymmetrically Tuned Filter

Second version of the image rejection filter was designed, primarily, to improve the insertion loss, but also to further reduce insertion loss. The filter was first designed as a parallel–coupled broadband filter. It is a third order filter, synchronously tuned, i.e. all three resonators have the same length. The characteristic impedance of each parallel–coupled pair is 50 Ohms.

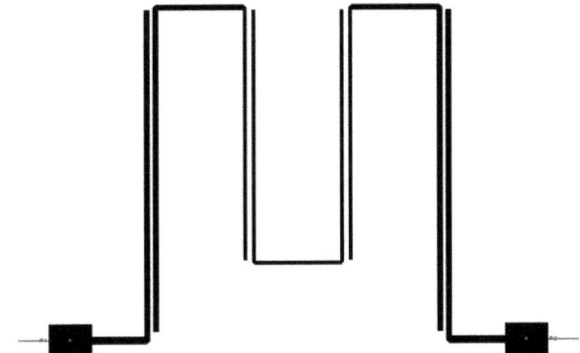

Figure 4. 34 Filter layout with ports for Momentum simulation.

Such a filter is very long and narrow, which is very unpractical for transmitter integration, so the filter shape was changed into meandering structure (see Figure 4.34). The parallel–coupled pairs are at 160 μm distance, to reduce coupling, which deteriorates selectivity. The filter was asynchronously tuned (the second resonator is shorter than the other two). The filter was optimized in such a way to have two transmission poles around 61.5 GHz, and the third pole is at the other end of the passband (see Fig 4.36). This improves both the insertion loss as well as image rejection. The characteristic impedance of the first and the last parallel–coupled pair should be kept at 50 Ohm for good return loss of the filter, whereas the characteristic impedance of the two middle pairs can vary significantly with little impact on the return loss.

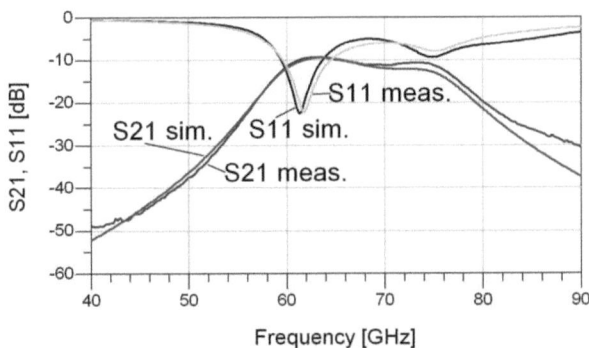

Figure 4. 35 Simulated and measured S parameters of the filter.

Figure 4.34 shows the layout of the filter for Momentum simulation. The first and the last parallel–coupled pair has width 10 μm and spacing 6 μm. The two middle pairs have width 6 μm and spacing 10 μm. The filter dimensions are 0.56×0.62 mm^2. Substrate definition is shown is Figure 4.29c. Simulation and measurement results match very well (see Figure 4.35). Insertion loss is 9.7 dB, image rejection 24.8 dB and return loss –22 dB with pads, and –13 dB without pads. Figure 4.36 shows simulated filter S parameters for the cases with and without ohmic losses, as well as with and without radiation (simulating option of Momentum). These simulations give us insight of effects of ohmic loss and radiation. It can be seen that most attenuation in the passband comes from the radiation. It also causes frequency shift of the passband.

This filter was used for the TX version II.

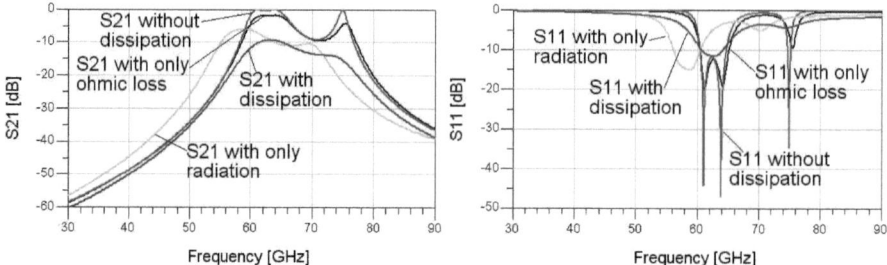

Figure 4. 36 Simulated S parameters with ohmic losses and radiation (i.e. with dissipation) and without.

4.4.4.3. Parallel–Coupled Asymmetrically Tuned Filter for the IEEE 802.15.3c Standard

The filter is intended for the IEEE 802.15.3c standard. Requirements for the filter are a large passband of 9 GHz (57–to–66 GHz) and 30 dB image rejection below 40 GHz. Selectivity requirements are not relaxed, but the filter needs good matching and low insertion loss in a large frequency range.

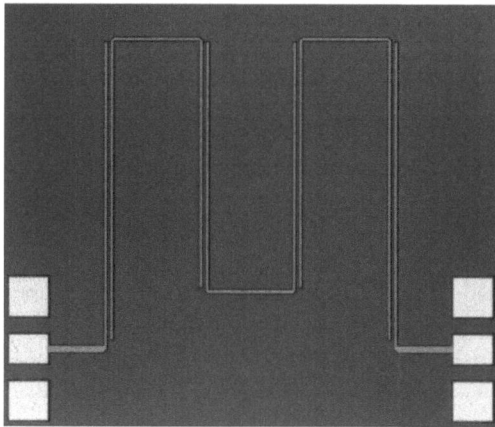

Figure 4. 37 Photo of the manufactured filter.

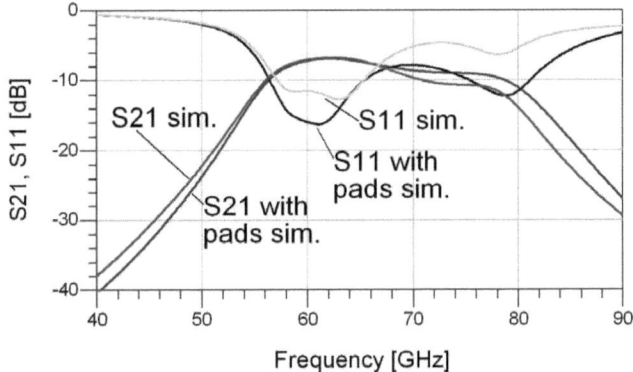

Figure 4.38 Simulated S parameters of the filter with and without pads.

The filter was designed and optimized in the same way as the filter in the previous section. The difference here is that the transmission poles are further apart, to cover the large bandwidth. Filter layout photo is shown in Figure 4.37. All four coupled pair have width 6 μm and spacing 6 μm. The filter dimensions are 0.55×0.60 mm^2. Substrate definition is shown in Figure 4.29c. Measured and simulated results are shown in Figure 4.37. They match well, but the transmission poles are closer in measurement than in simulation. Insertion loss is between 6.7 and 8.2 dB inside the range: 57- to–66 GHz. The minimum image rejection is 38 dB. The return loss is from –13 to –6 dB with pads. The simulated return loss without pads is from –13 to –9 dB. The effect of pads on the simulated S parameters can be seen in Figure 4.38.

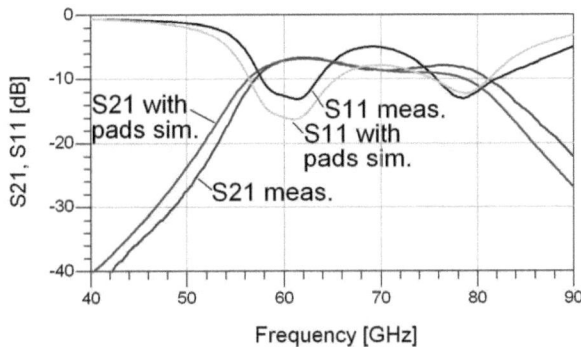

Figure 4.39 Measured and simulated S parameters of the filter with pads.

4.4.4.4. Lumped Element Filter for the IEEE 802.15.3c Standard

Filters with lumped elements have considerably lower selectivity than with microstrip resonators. Since selectivity requirements for the IEEE 802.15.3c standard are very low, lumped element filter is a good candidate for realization of this kind of filter. Similar approach was used in [4.17] for a 77 GHz filter.

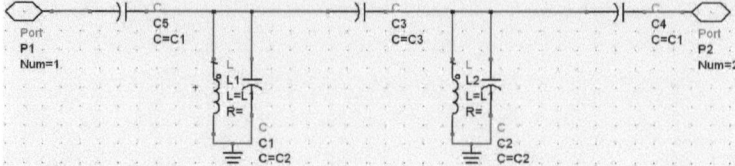

Figure 4. 40 Schematic of the lumped element filter.

The filter was realized with two transmission poles. Lumped model schematic is shown in Figure 4.40. Filter simulation included Momentum simulation of all transmission lines with pads (see Figure 4.41). Based on this simulation, a component was created and it was used in schematic with capacitors for the filter simulation. Capacitors have value: $C_1 = 27$ fF, $C_2 = 45$ fF and $C_3 = 20$ fF. The filter has poles in simulation at 58.5 GHz and 65 GHz. 3 dB bandwidth is 17 GHz (54-to-77 GHz). Insertion loss in the band of interest (57-to-66 GHz) is 2.8 to 3.8 dB. Return loss is below −15 dB in the whole range. The minimum image rejection is 25 dB (see Figure 4.42). The filter size without pads is only 220×90 μm². Small size, low insertion loss and good return loss make this filter better solution for this application than the microstrip filter presented in section 4.4.4.3. Measured S parameters match well with the simulation (Figure 4.42). The insertion loss is from 3.3 to 4.3 dB (57-to-66 GHz) and the minimum image rejection is 23 dB. The upper transmission pole is shifted by 3 GHz. It can be shifted back for better return loss.

This filter was used for the TX version III. In addition to the image–rejection, the filter also attenuates the VCO feed-thru at 48 GHz. This is important because the TX and RX antennas are in close proximity (see Figure 6. 11) and a strong VCO feed-thru could saturate the LNA.

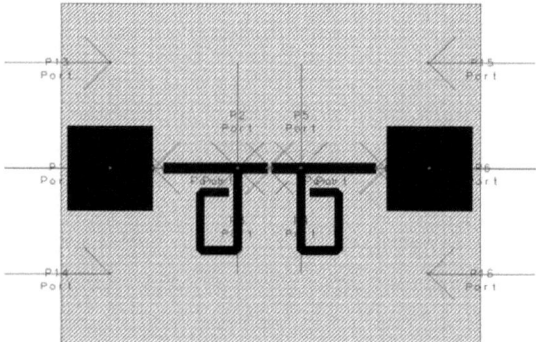

Figure 4. 41 Filter layout used for Momentum simulation.

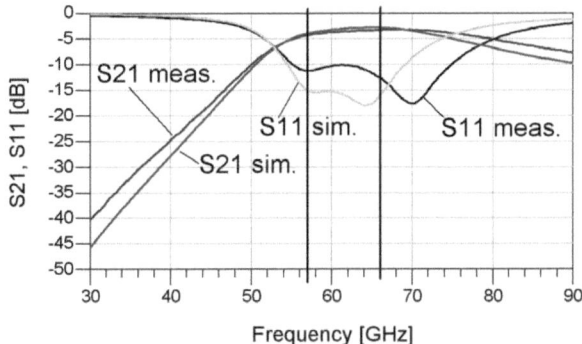

Figure 4. 42 Measured and simulated S parameters of the lumped filter.

4.4.4.5. Narrowband on–Board Filter

Material used for the board is Rogers 3003 with relative electric permittivity $\varepsilon_r = 3$ and thickness of 5 mil. This is too thin for mechanical stability, so additional layer of FR4 was used. Metal layers are from copper, with additional gilding of the upper and lower copper layer. Top metal layer has thickness of 35 μm. Gilding increases conductivity and reduces filter insertion loss. Design rules for the PCB specify that minimum line width is 100 μm and minimum metal separation is 110 μm. These rules are stricter than for some other PCB production technologies, but, on the other hand, price is lower.

Design goals for the filter are: centre frequency at 61 GHz with 2 GHz passband bandwidth; more than 20 dB image suppression at 51 GHz (with respect to 61 GHz), small insertion loss (3–to–4 dB), compact design and a design robust to production tolerances. Tolerances are relatively high since the used inexpensive technology is not very accurate.

Figure 4.43 shows produced filter photo. It features a simple layout composed of three coupling structures: two with side coupling and one with end coupling. Needed image rejection of 20 dB can be easily achieved with this structure and the design is compact.

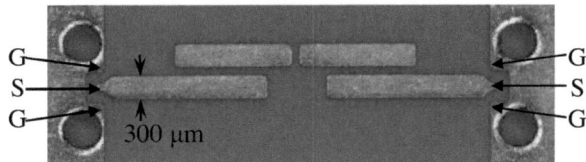

Figure 4. 43 Simulated bandpass filter S parameters with and without dissipation.

Filter is implemented using 50 Ω transmission lines (width 300 μm). Using narrower lines enables better design with respect both to insertion loss (due to stronger coupling) and selectivity. However, filter characteristics vary significantly due to process tolerances if narrow (100 μm) transmission lines are used.

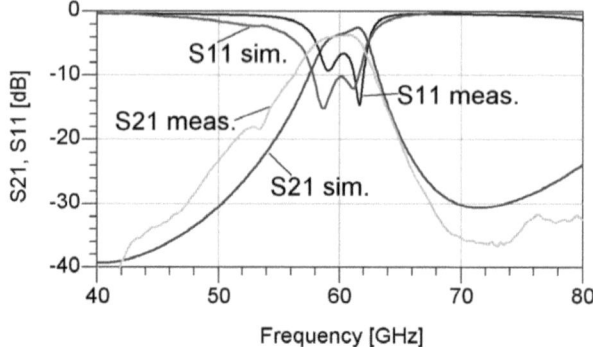

Figure 4.44 Simulated and measured S parameters of the on–board filter.

Filter design had to include structure for the measurements. The filter was measured probing on a wafer probe. Probes with ground–signal–ground (GSG) footprint and 250 μm pitch were used. Probing position is indicated on the Figure 4.43 with GSG letters. EM simulation showed that the probing structure has very good matching to a 50 Ω line.

Comparison of simulated and measured results is show in Figure 4.44. The insertion loss of the filter with the probing structure is 3.8 dB (both in simulation and measurements). De–embedded value is 3.5 dB. The measured characteristic is more flat for the passband. The image–rejection is 26 dB in simulation and 20 dB in measurement, which is not optimal but is sufficient. The measured results over 10 boards for the insertion loss vary 0.3 dB and for the image–rejection 1.5 dB. These tolerances are acceptable.

This filter was used for a test version of a TX without a PA and an integrated filter.

4.5. Summary

In this chapter the design of the integrated image–rejection filters has been analyzed. There is limited work done on filters for 60 GHz applications, and filters which were published were produced on–board. The analysis presented here is the first on integrated filters for 60 GHz.

The main problems related to the design of integrated filters arise from the low quality factor Q of the integrated resonators. The loss comes mainly from the radiation and the ohmic loss. Bandpass filters are much more susceptible to the low Q, especially the narrowband filters.

Two measures to reduce the insertion loss of the image–rejection filters were suggested. One is to design the filter as broadband. The lowest transmission pole should match the signal frequency. This measure deteriorates selectivity, so the minimum required image–rejection will limit the width of the passband.

The second measure is to optimize the filter asynchronously. The filter was designed so that two transmission poles are around the signal frequency. This measure will improve both the insertion loss and the image–rejection.

In the case when the image is far away from the signal (AFE version II) and when the filter requirement are not strict, the best option is to design a lumped element integrated filter. Lumped element filters are very small and can be optimized for very good insertion loss and return loss.

The on–board filters can be designed for a strict set of specifications. The board production process must, however, have very strict tolerances. The on–board filter presented here was not designed for strict specifications. Hence, it was produced on a low–cost PCB. The filter has a robust, compact design.

Chapter 5

Power Amplifier

5.1. Introduction

This chapter presents the design of the power amplifier, which is one of the most important blocks in RF transmitters. The first section addresses the basic concepts related to the PA design, such as amplifier stability, gain, input/output matching, saturated and linear output power, efficiency, power dissipation, small-signal and large-signal matching and nonlinear distortion. Main power combining techniques and their limits are also discussed.

Modelling of passive structures, and the test structures used to verify EM simulations and measurement procedure are presented in the third section. The procedure of the PA design for our 60 GHz wireless communications system is presented in the next section. The choice of topology, class of operation and power combining technique is discussed. The difficulties of making the layout from the schematic for mm-wave circuits are explained.The measurement results for the fabricated PA are presented along with a comparison of the state-of-the-art power amplifier designs at 60 GHz in SiGe an CMOS technologies.

PA theory application field is broad, but the discussions in this chapter will be focused on the areas relevant for 60 GHz frequency range and SiGe HBT transistors. For example, the achievable output power in SiGe at 60 GHz is around 20 dBm, which is much lower than output power levels at lower frequencies – above 30 dBm. Consequently, the PA design won't deal with extreme problems of self heating common to low frequency high power PAs.

5.2. Power Amplifier Theory

Design of a power amplifier entails achieving different and often conflicting requirements regarding its operation characteristics, such as output power, gain, stability of operation, input and output matching, power-added efficiency etc. Before discussing operation of an amplifier with high output power, i.e. power amplifier, let us first see basic design equations and stability analysis of a small-signal amplifier. This analysis is, in the most part, applicable on the power amplifiers.

The block diagram of a simple, one-stage, microwave amplifier is shown in Figure 5. 1. It consists of a transistor (BJT, HBT or FET), which is characterized by its S-parameters. The transistor is biased using biasing circuitry, which is for simplicity not shown in the picture. The transistor input and output impedance is matched to the source and load using the input and output matching networks. These networks are passive, while biasing circuitry is often active.

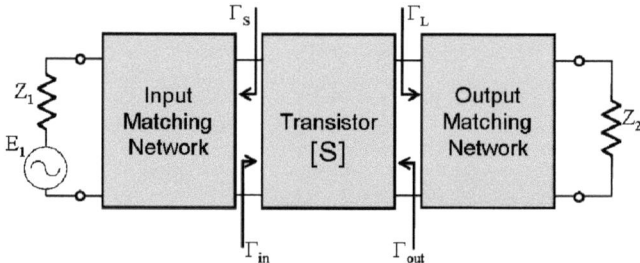

Figure 5. 1 A block diagram of a one-stage microwave amplifier.

Transistor S-parameters are frequency dependant and the following analysis is accurate only at the frequency of S-parameters, but practically it can be seen as a narrow bandwidth analysis because S-parameters of the transistor are approximately constant in a small frequency range.

5.2.1. Stability Considerations of an Amplifier

The first step in an amplifier design is to check if the transistor is unconditionally stable, i.e. if it would oscillate for certain impedances at the input and/or output. This is important because an amplifier oscillation is a highly undesirable phenomenon. In such cases the amplifier performance may change strongly and it may lead to circuitry damage. The oscillation as an unwanted, usually strong, signal represents noise to the useful signal and in transmitters this signals could lead to transmission in a forbidden band.

For an amplifier as presented in Figure 5. 1, oscillations are possible when either the input or output port presents a negative resistance. This can be mathematically expressed using the input and output reflection coefficient as: $|\Gamma_{in}| > 1$ or $|\Gamma_{out}| > 1$. Γ_{in} and Γ_{out} can be expressed in terms of transistor S-parameters and source and load reflection coefficients (Γ_S and Γ_L) as given [5.1]:

$$\Gamma_{in} = s_{11} + \frac{s_{12}s_{21}\Gamma_L}{1 - s_{22}\Gamma_L} \tag{5.1}$$

$$\Gamma_{out} = s_{22} + \frac{s_{12}s_{21}\Gamma_S}{1 - s_{11}\Gamma_S} \tag{5.2}$$

The input and output port impedances Z_1 and Z_2 (Figure 5.1) have positive resistance (standard 50 Ohm). After transformation with passive matching networks the real part of the impedance in every case remains positive, i.e. $|\Gamma_S| < 1$ or $|\Gamma_L| < 1$. If an amplifier is stable for any impedance (with positive resistance) at the input and output port (i.e. for any $|\Gamma_S| < 1$ and $|\Gamma_L| < 1$) for any given frequency, we say that it is unconditionally stable. If this is not the case for some input or output impedances, the amplifier is potentially unstable.

When analyzing a potentially unstable amplifier it is useful and common to use the graphical representation of the reflection coefficients (Γ_S and Γ_L) in the Smith chart. Solving equations $|\Gamma_{in}| = 1$ and $|\Gamma_{out}| = 1$ for Γ_L and Γ_S gives output and input stability circles in the Smith chart, respectively. The next step is to determine whether the stable region is inside or outside of the circle. A simple way to do that is to take the point in the centre of the circle ($|\Gamma_L| = 0$) and if it is true that $|\Gamma_{in}| = |s_{11}| < 1$ then the centre of the Smith chart is in the stable region, and vice versa. Analog is true for the input stability circles.

If it is possible, the goal is to design an unconditionally stable amplifier. This is checked using stability criteria derived from the condition that the stable regions of the input and output stability circles cover the whole Smith chart. Mathematical representation of necessary and sufficient conditions for unconditional stability is as follows [5.1]:

$$K = \frac{1 - |s_{11}|^2 - |s_{22}|^2 + |\Delta|^2}{2|s_{12}s_{21}|} > 1 \tag{5.3}$$

$$|\Delta| < 1 \tag{5.4}$$

where,

$$\Delta = s_{11}s_{22} - s_{12}s_{21} \tag{5.5}$$

Another way to mathematically express the necessary and sufficient conditions for unconditional stability is:

$$K > 1 \tag{5.6}$$

$$B_1 = 1 + |s_{11}|^2 - |s_{22}|^2 - |\Delta|^2 < 0 \tag{5.7}$$

Other stability factors are commonly used in the literature such as the μ-parameter [5.1].

Some microwave transistors are unconditionally stable, but some are not. If 0<K<1 for most of the impedances in the Smith chart the amplifier would be stable, but for −1<K<0 most of the impedances in the Smith chart would cause the amplifier to oscillate.

As an illustration we can see the stability factor of an HBT npn201_8 transistor in IHP SiGe H1 technology in a common emitter configuration, shown in Figure 5. 2a. For frequencies below 40 GHz the transistor is not unconditionally stable ($K<1$). The graphical representation of the impedances at the input and output is shown in Figure 5. 2b at 20 GHz. For all input impedances inside the blue (unity) circle and outside of the red (source stability) circle the transistor would be stable (at 20 GHz). For all output impedances inside the blue (unity) circle and outside of the purple (load stability) circle the transistor would be stable (at 20 GHz).

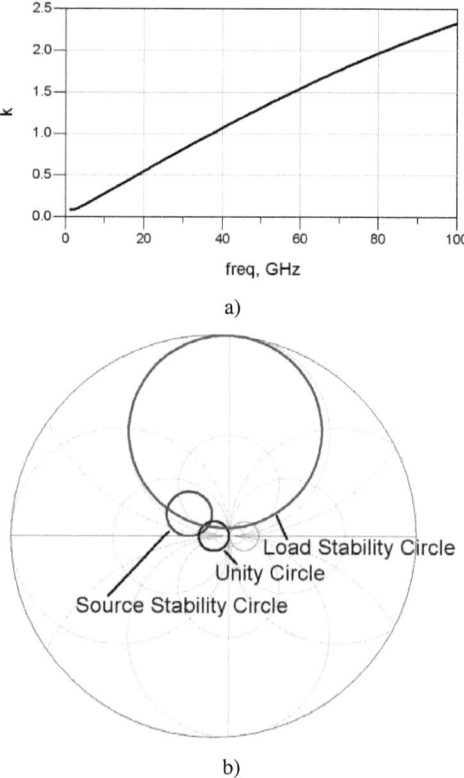

Figure 5. 2 a) Stability factor for a HBT npn201_8 transistor in a common emitter configuration. b) The source and load stability circles at 20 GHz ($K \cong 0.6$) with the stability region outside of the circles. The unity circle is $|\Gamma| = 1$.

5.2.2. Small-Signal Operation of an Amplifier

There are several power gain definitions in the literature, which are used in the design of microwave amplifiers. Let us first see the powers used in their definitions. P_L represents the power delivered to the load (Z_2 in Figure 5. 1). P_{AVN} represents the power available from the (output matching) network. If the amplifier is ideally matched to the load impedance then $P_L = P_{AVN}$. P_{IN} represents the power delivered to the amplifier from the source. P_{AVS} represents the power available from the source. If the amplifier is ideally matched to the source impedance then $P_{IN} = P_{AVS}$. Based on these power definitions, the following power gain definitions are used: the transducer gain G_T, the power gain G_p and the available power gain G_A. They are defined as [5.1]:

$$G_T = \frac{P_L}{P_{AVS}} = \frac{1-|\Gamma_S|^2}{|1-\Gamma_{in}\Gamma_S|^2}|s_{21}|^2\frac{1-|\Gamma_L|^2}{|1-s_{22}\Gamma_L|^2} \tag{5.8}$$

$$G_p = \frac{P_L}{P_{in}} = \frac{1}{1-|\Gamma_{in}|^2}|s_{21}|^2\frac{1-|\Gamma_L|^2}{|1-s_{22}\Gamma_L|^2} \tag{5.9}$$

$$G_A = \frac{P_{AVN}}{P_{AVS}} = \frac{1-|\Gamma_S|^2}{|1-s_{11}\Gamma_S|^2}|s_{21}|^2\frac{1}{1-|\Gamma_{out}|^2} \tag{5.10}$$

If the amplifier is ideally matched to the load (output conjugate match $\Gamma_{out} = \Gamma_L^*$) than $G_T = G_p$, and if it is at the same time ideally matched to the source (input conjugate match $\Gamma_{in} = \Gamma_S^*$) then $G_{Tmax} = G_{pmax} = G_{Amax}$. This is also called the simultaneous conjugate match, and it gives the maximum gain for the amplifier, when it is unconditionally stable. For the conditionally stable case G_{Tmax} is given as [5.1]:

$$G_{T\max} = \left|\frac{s_{21}}{s_{12}}\right|\left(K - \sqrt{K^2 - 1}\right) \tag{5.11}$$

and for K=1 we get the maximum stable gain $G_{T\max} = |s_{21}/s_{12}|$.

From the transistor S-parameters, the stability and Γ_S and Γ_L for the maximum gain can be calculated [5.1]. If the transistor is unconditionally stable, input and output matching networks to transform port impedances Z_1 and Z_2 to the required values for the transistor can be designed. If the transistor is conditionally stable, additional circuitry can be added [5.1] to make it unconditionally stable – which is usually preferable, or if the reflection coefficients Γ_S and Γ_L are in the stable region – the amplifier can be left as conditionally stable. We should note here that adding additional circuitry for stability usually degrades amplifier performance in terms of amplification or output power.

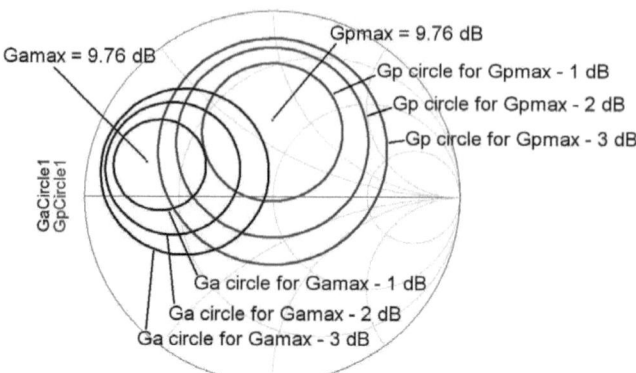

Figure 5. 3 The Smith chart with constant gain G_A and G_p circles for an HBT npn201_8 transistor in IHP H1 technology in a common emitter configuration at 60 GHz.

Amplifier design uses graphical representation in the Smith chart to present the optimal values for G_p and G_A. It can be shown that point for a constant G_p or G_A in the Smith chart form circles [5.1]. Equations for the centre and radius of these constant-gain circles can be found in [5.1].

An example of constant-gain G_p or G_A circles is shown in Figure 5. 3, depicting points for G_{pmax} and G_{Amax} and circles with 1, 2 and 3 dB lower amplification. The circles are calculated for an HBT npn201_8 transistor in IHP H1 technology in a common emitter configuration at 60 GHz.

As shown in the section 5.2.1, the transistor is stable at 60 GHz, but not below 40 GHz. A structure, similar to the one shown in Figure 5. 14, can be added to make the transistor unconditionally stable without significant effect on the performance at 60 GHz.

5.2.3. Amplifier Linearity

The analysis so far, has assumed transistor S-parameters that are not dependant on the level of the input signal. This is of course not realistic, because it would mean that the amplifier can provide unlimited power at the output. The output power is usually limited by the transistor breakdown voltages and the supply voltage, and very rarely by the maximum allowed current.

Figure 5. 4 shows the principal of amplification with a bipolar transistor in common emitter configuration. The figure shows transfer characteristic of the input signal (base-emitter voltage) to the output signal (collector current) and the quiescent dc operating point of the transistor. A relatively small and a relatively large signal are amplified. It can be seen that the large sinusoidal signal gets distorted during amplification and the small signal doesn't. This is because the transfer characteristic is not linear, but in a small range around the quiescent point it approximately is. If we assume that the amplifier is a memoryless system (in most cases true) the transfer characteristic is given as [5.2]:

$$y_o(t) = a_1 x_i(t) + a_2 x_i^2(t) + a_3 x_i^3(t) + a_4 x_i^4(t) + \ldots \tag{5.12}$$

but it is usually approximated with the first three elements of the series.

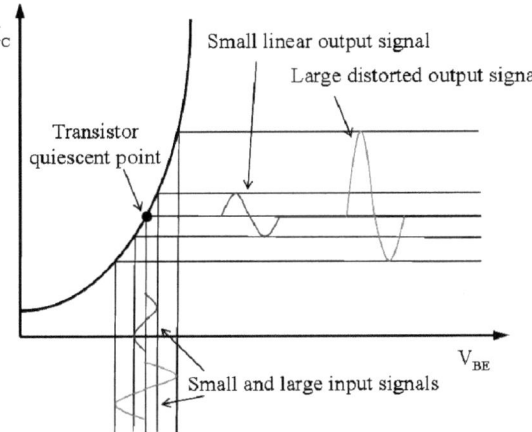

Figure 5.4 Amplification and distortion of a small and large signal (voltage to current) in a bipolar transistor with common emitter configuration.

Output signal distortion is possible both in the amplitude (AM-AM distortion) and in the phase (AM-PM distortion). Amplifier nonlinearity causes different phenomena, such as:

- creation of higher harmonics from the sinusoidal input signal,

- gain compression (AM-AM distortion),

- desensitization and blocking of a weak desired signal in the presence of a strong interferer (common for receivers),

- transfer of the amplitude modulation from a strong interferer to a weak desired signal (called cross-modulation, common in receivers), and

- creation of intermodulation (IM) products from two signals of different frequency.

Since power amplifiers don't have the problem of strong interferer, desensitization, blocking and cross-modulation phenomena are not of interest in a PA analysis. Other nonlinear phenomena will be analyzed in the following subsections.

5.2.3.1. AM-AM Distortion and 1dB Compression Point

For a sinusoidal input $x_i(t) = A_1 \cos(\omega_1 t)$, the output signal is [5.2]:

$$y_o(t) = \frac{a_2 A_1^2}{2} + \left(a_1 A_1 + \frac{3 a_3 A_1^3}{4}\right)\cos(\omega_1 t) + \frac{a_2 A_1^2}{2}\cos(2\omega_1 t) + \frac{a_3 A_1^3}{4}\cos(3\omega_1 t) \tag{5.13}$$

It can be seen that the output contains harmonics of the input signal, i.e. signals with frequencies which are integer multiples of the input signal frequency.

The amplification of the input signal is $a_1 + 3a_3 A_1^2/4$. For small A_1 the amplification (a_1) is approximately constant. For A_1 large enough, the amplification drops, because $a_3 < 0$ ($a_3 > 0$ would mean that the amplifier oscillates, or that the quiescent point is close to the breakdown voltage). Typical shape of the P_{out} vs. P_{in} curve in the log-log scale is shown in Figure 5.5.

1dB compression point (P1dB) is an often used parameter, which is a figure of merit for amplifier linearity. Output referred P1dB represents the power of the amplified signal at the output, when the amplification drops by 1 dB (see Figure 5.5). The power of the input signal represents the input referred P1dB. Further in the text, P1dB represents output referred value, unless otherwise stated.

It should be noted that P1dB is only a figure of merit for the AM-AM distortion. It doesn't tell us anything about the AM-PM distortion. A PA can have low AM-AM, but high AM-PM distortion. However, in many cases, these values are strongly correlated and P1dB is used as an indicator for the AM-PM distortion, too.

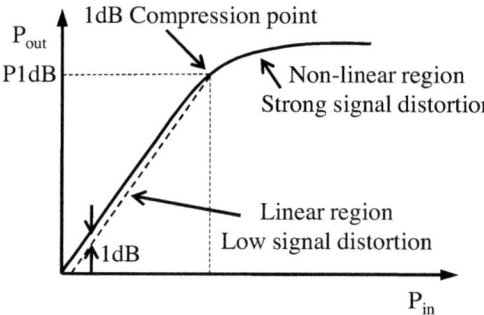

Figure 5.5 Typical P_{out} vs. P_{in} curve showing the 1dB compression point and regions of low and high distortion.

5.2.3.2. AM-PM Distortion

AM-PM distortion is a phenomenon that the phase of the output signal depends on the level of the input signal. This type of distortion is important in systems that employ some kind of phase modulation. It is a result of a complex dependency of transistor parameters, such as base-collector capacitance, on the level of the input signal [5.3]. Typical measurement of the relative phase versus the input power (log scale) is shown in Figure 5.6. The best performance exhibit amplifiers in class A (high-linearity PA in the picture). The relative phase changes only for power above P1dB. Class AB amplifiers (low-linearity PA in the picture) have significant change of the relative phase even for very low levels of the input signals, much below P1dB. The unit for the AM-PM distortion is deg/dB.

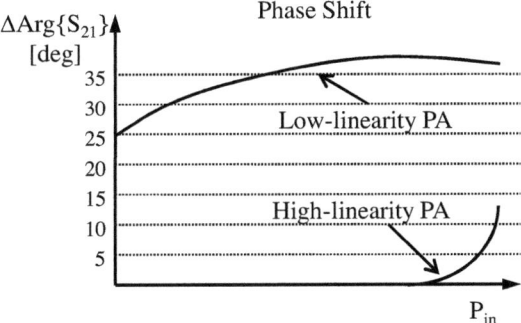

Figure 5. 6 Typical measurement of the relative phase for a high-linearity PA (class A) and low-linearity PA (class AB). P_{in} is in log scale.

5.2.3.3. Intermodulation Distortion

Intermodulation (IM) distortion appears when there are at least two tones at the input, i.e. for $x_i(t) = A_1 \cos(\omega_1 t) + A_2 \cos(\omega 2t)$. The output contains these signals (frequencies ω_1, ω_2), their harmonics (frequencies $2\omega_1$, $3\omega_1$, ... $2\omega_2$, $3\omega_2$, ...) and the result of mixing of the input tones (frequencies $\omega_1 \pm \omega_2$, $2\omega_1 \pm \omega_2$, $\omega_1 \pm 2\omega_2$, ...). The amplitude of each component can be calculated by submitting this input signal into the transfer characteristic equation (5.12) as given in [5.2].

Communication systems are usually more sensitive to IM components, than signal harmonics, because they usually fall far away from the used spectrum and are easy to filter out. Third order IM component $2\omega_1-\omega_2$, and $\omega_1-2\omega_2$ (see Figure 5. 7) for close frequencies ω_1 and ω_2, fall inside the spectrum and can't be filtered. These IM components cause so called spectrum regrowth phenomenon.

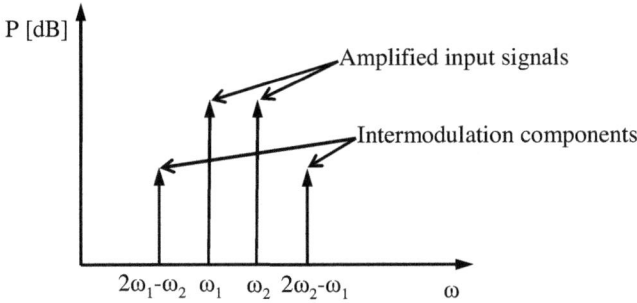

Figure 5. 7 Output spectrum of an amplifier around the input tones.

The figure of merit for IM distortion is called third order intercept point (IP$_3$). IIP$_3$ is input referred IP$_3$ and OIP$_3$ is output referred IP$_3$. It can be shown that as a rule of thumb the input referred P1dB and IIP$_3$ have the following relation [5.2]:

$$IIP_3 = P1dB + 9.6dB \tag{5.14}$$

5.2.4. Linearity of Cascaded Amplifiers

If one stage amplifier can't provide sufficient gain, the common solution is to cascade amplifiers. The total gain (G_{tot}) of a cascade of n amplifiers is the product of gain of each amplifier (G_i), i.e. the sum of gains in dB [5.2]:

$$G_{tot} = \prod_{i=1}^{n} G_i \tag{5.15}$$

Cascading amplifiers has an adverse effect on the overall linearity. If i^{th} amplifier has gain G_i and input referred IP$_3$ IIP$_{3,i}$, total IIP$_3$ (IIP$_{3,tot}$) is approximately given as [5.2]:

$$\frac{1}{IIP_{3,tot}^2} = \frac{1}{IIP_{3,1}^2} + \frac{G_1}{IIP_{3,2}^2} + \ldots + \frac{G_1 G_2 \cdots G_{n-2}}{IIP_{3,n-1}^2} + \frac{G_1 G_2 \cdots G_{n-1}}{IIP_{3,n}^2} \tag{5.16}$$

The equation (5.16) shows that all stages contribute to the total IIP$_3$, but the last stage is scaled by the gain of all the previous stages – making it, usually, the dominant factor. For a large G_n, the effect of previous stages is negligible. For a small G_n, the contribution of previous stages may even dominate.

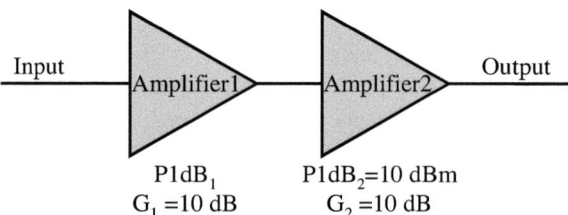

Figure 5. 8 Two cascaded amplifiers with equal gain.

ΔP1dB [dB]	10	7	5	3	0
P1dB$_{tot}$–P1dB$_2$ [dB] for two stages	−2.59	−1.38	−0.86	−0.56	−0.28
P1dB$_{tot}$–P1dB$_2$ [dB] for three stages	−4.2	−1.94	−1.13	−0.66	−0.3

Table 5. I Degradation of output P1dB due to amplifier cascading for two and three stages

Using multiple stages in the mm-wave range is often necessary because of the relatively low gain that can be achieved per stage. It is of interest to see the effect of cascading amplifiers with moderate gain on the total output P1dB (P1dB$_{tot}$). This effect can be calculated from equations (5.16) and (5.14). We will examine simple cases of two (shown in Figure 5.8) and three amplifiers, with equal (relatively low) gain of 10 dB. The last stage has output referred P1dB of 10 dBm, and all three amplifiers have equal gain of 10 dB. Table 5.I presents the reduction of P1dB$_{tot}$ for different values of ΔP1dB (ΔP1dB = P1dB$_2$–P1dB$_1$ = P1dB$_3$–P1dB$_2$). As can be seen from Table 5.I, for ΔP1dB=10 dB (i.e. P1dB$_2$–P1dB$_1$ = G$_2$), P1dB$_{tot}$ is reduced by 2.59 dB. This is for most applications too large reduction. Even for ΔP1dB of just 3 dB, the reduction is almost 0.6 dB, which is for a mm-wave PA significant due to limited achievable values and problematic power combining (see section 5.2.7). For three stages the need for small ΔP1dB is even greater. This means that even the earlier stages of a PA need larger transistors and higher power consumption, causing more self heating and reducing the overall efficiency.

5.2.5. Efficiency and PA Classes

PA efficiency, denoted as η, is defined as a ratio of the delivered RF power (P$_{RFout}$) to the load and the dissipated dc power by the PA:

$$\eta = \frac{P_{RFout}}{P_{DC}} \qquad (5.17)$$

An ideal PA has $\eta = 1$, i.e. 100%. One more figure of merit commonly used for PA efficiency is power-added efficiency (PAE), which represents the ratio of the delivered RF power reduced by the input RF power (P$_{RFin}$) and the consumed dc power:

$$PAE = \frac{P_{RFout} - P_{RFin}}{P_{DC}} = \frac{P_{RFout}}{P_{DC}}\left(1 + \frac{1}{G_T}\right) \qquad (5.18)$$

PAE is more commonly used for high-frequency PAs than η, because it gives a fairer comparison between different PAs. Two PAs with equal P$_{RFout}$ and dc power consumption and different gain have the same η, but since more gain requires more dc power, the PA with higher gain is really more efficient, and PAE unlike η will show this.

PAs have historically been divided into several classes: A, AB, B, C for analog input signals and D, E (and others) for digital input signals [5.2]. Classes refer to the topology and biasing of the transistor, and can achieve different levels of efficiency. Figure 5.9 shows typical IV curves of a BJT transistor, and load lines with quiescent points for the analog A, AB, B and C classes. Class A is the most linear, i.e. the largest P1dB with the lowest phase distortion can be achieved in this class. However, it is the least efficient with theoretical maximum for efficiency of 50% [5.2]. The main disadvantage is that the power dissipation is high and constant even when there is no signal at the input. Classes B and C have higher theoretical maximum for efficiency of 78% and 90%, respectively [5.2]. They also have no power dissipation when there is no input present, and class AB has low power dissipation. However, they have high distortion, because the signal is not present for the full period of the input signal (see Figure 5.9).

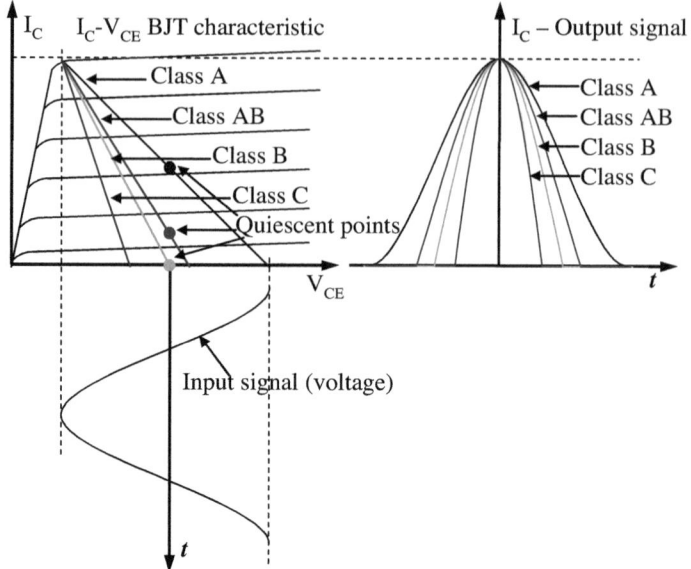

Figure 5. 9 Load lines and quiescent points for different PA classes of operation with the sinusoidal input signal and resulting output waveforms.

5.2.6. Load-Pull Measurements

Section 5.2.2 showed amplifier design for maximum gain for small signals. For a power amplifier, the main parameter of interest is not gain but output power (P1dB or P_{sat}) and PA design is centred around output power optimization. This, however, can't be done with small signal S-parameters, because they depend on the power level. Load-pull measurement is used to test the transistor output power versus the complex load seen by the transistor [5.2]. Figure 5. 10 shows the measurement setup. The tuner behind the transistor creates different load impedances Z_L and the power meter measures the delivered power. When Z_L changes, so does the input impedance Z_{in}. The tuner in front of the transistor is used to match the changing Z_{in} to the signal generator, so that the transistor always gets the same power at the input. In such a way, the measured output power depends only on Z_L.

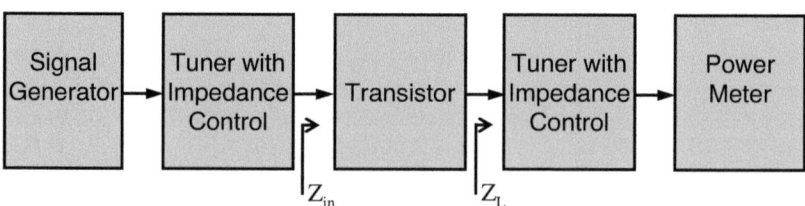

Figure 5. 10 Load-pull measurement setup.

The load pull measurement system is usually automatic and produces in the Smith chart the optimum output impedance and curves of constant output power (see Figure 5.15b). These systems are, however, relatively complex and not available for the mm-wave frequencies, because of difficulty to make tuners for high frequencies. For this range one has to rely on the transistor models, which are not accurate for high power levels. For class A PAs and P1dB optimization (i.e. for low nonlinearities) transistor models are usually sufficient [5.3].

5.2.7. Power Combining Techniques

Power combining is used when one transistor cannot supply sufficient power level at the output. There are two main techniques which are commonly used. One is to parallelize several transistors as shown in Figure 5.11a. The other is to use hybrid dividers to equally split the input signal power and feed it to two or more amplifiers (see Figure 5.11b) and then to sum the power of the output signals by means of hybrid combiners.

The first technique is simple for implementation but has important disadvantages:

1. The input and output impedance of n parallel transistors is n times smaller, and for n large enough, it becomes comparable to the resistance of the input or output matching network. The loss in the matching network becomes large and overall efficiency and delivered output power drops. This means that paralleling transistors makes sense only for a relatively small number of transistors.

2. If one transistor fails, the whole amplifier fails.

3. All transistors should be well matched and identical in order to keep equal load sharing. The last problem is easier to solve for the integrated as opposed to the discrete case. If the transistors are placed one next to the other in the layout, their parameters will be very similar resulting in good load sharing.

The technique using hybrid dividers and couplers is advantageous because it doesn't suffer from the problems of the parallelizing technique. Here, if one transistor fails, the PA will continue to work with lower output power. But this technique has disadvantages too:

1. The whole amplifier with input and output matching networks and biasing circuitry is copied, not just transistors. This results in much larger PA area, which is an expensive resource for an integrated solution.

2. The area is further increased with the use of hybrid combiners, usually Wilkinson couplers. For the lower frequencies, the size of hybrid combiners is large, which is a problem for an integrated solution. For the higher frequencies, the size is smaller, but the insertion loss is significant, reducing the output power of the PA.

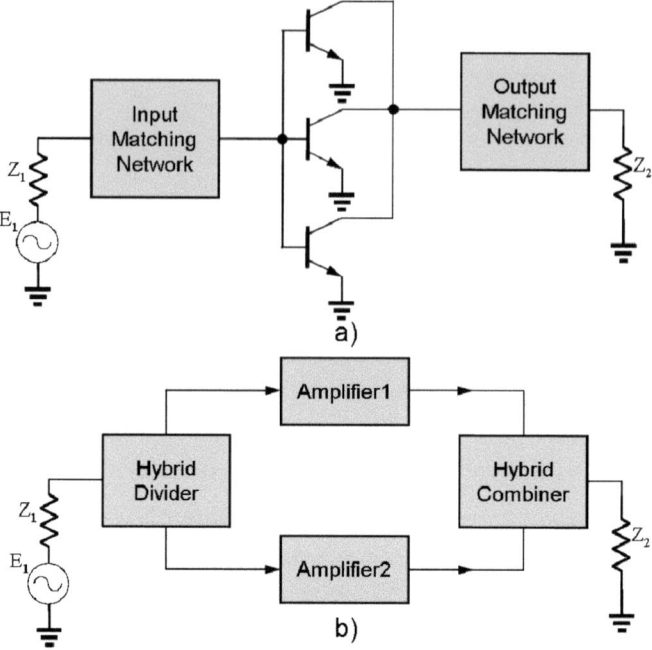

Figure 5.11 Power combining techniques: Parallelizing transistors a) and using hybrid dividers to split input signal and hybrid combiners to add output signals.

We can summarize that for the mm-wave range, integrated PAs power combining is possible and efficient for a very limited number of transistors/amplifiers. Another approach in power combining is to use multiple PAs and multiple antennas as in beam forming architecture. The power is then "combined" in the air – without any loss.

5.3. Modeling of Distributed Passive Elements for Matching

Passive structures such as inductors, transmission lines (TLs) and metal lines are used for matching structures and as interconnect lines. In the microwave range these structures need to be properly modelled. Some of these structures are characterised in the literature with a closed form expressions, such as TLs and inductors, but EM simulation of the distributed passive elements is still extremely important. The EM simulation enables characterisation of the exact layout and calculates all parasitics and couplings, which can be very significant in the microwave range. Furthermore, many structures are not characterised by equations, and need to be simulated, for example metal lines without ground layer, or inverted TLs with top metal layer as ground and silicon below the signal line. Interconnect lines, even when very short ($10\,\mu$m), need to be simulated.

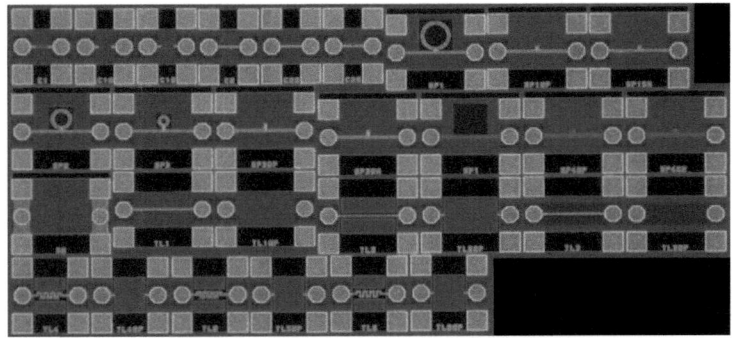

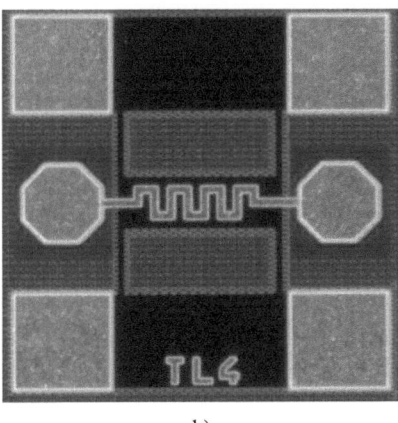

Figure 5. 12 a) Micrograph of the test structures b) Micrograph of the meandered TL.

The program used here to perform EM simulations is Momentum, which is integrated in Agilent ADS simulation software. It is a 2.5D simulator, which means that it doesn't take into account fields around structures in the vertical direction (i.e. vias). This is, however, not significant for an integrated design, due to the planar nature of the layout. After the simulation, the program allows simple creation of a schematic data component with the attached simulated S-parameter file. The component can be used in the schematic for any frequency domain simulation and in time domain after calculation of impulse responses for each port.

The program was first verified by comparing simulated and measured S parameters of a set of passive structures, which included TLs, meandered TLs, metal lines without ground layer etc. The comparison of the simulated and measured results is important to verify simulation setup, i.e. substrate definition, simulation ports, mesh, but also the measurement setup and the deembedding procedure.

Momentum allows description of planar substrates only. This is a problem for the top metal layer (TM2) for IHP technology, because the top SiO_2 layer is not flat but simply covers the top metal layer (see Figure 4. 29a). In such cases substrate definition is modified and presented as planer (Figure 4. 29b). This is not a problem for the PA design, because there are no structures with side coupling, as is the case for the image rejection filter.

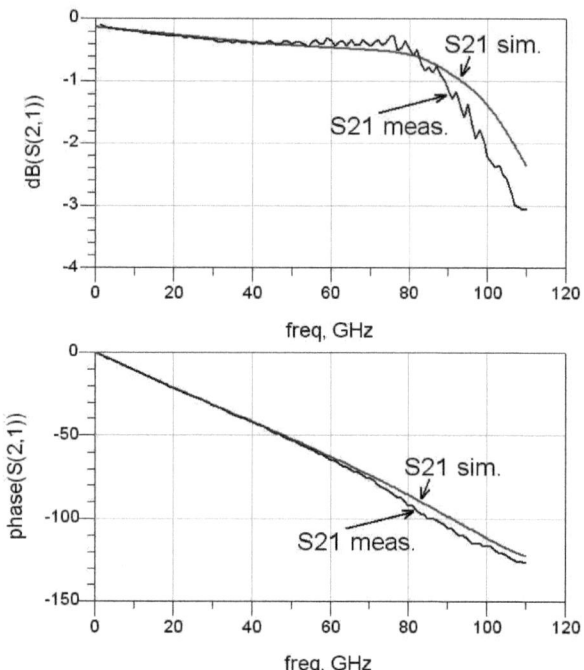

Figure 5. 13 Comparison of measured and simulated S_{21} parameter of the meandered line in the TM1.

Layout of the fabricated test structures is shown in Figure 5. 12a. Layout of the meandered TL4 is shown in Figure 5. 12b. The line is in Top Metal1 layer and ground layer in Metal1. The line is app. 400 μm long and 7 μm wide (50 Ohm TL).

Figure 5. 13 shows the comparison of the measured and simulated S_{21} parameter from dc to 110 GHz. We can see that the simulation results match well with the measurement and very well up to 60 GHz, which was the most important for the application. Other structures show good matching as well.

5.4. Power Amplifier Design in the 60 GHz Range

5.4.1. PA System Requirements

The required PA performance was analyzed in the second chapter. The conclusions are here restated. The PA has to provide high linear output power, at least 10 dBm, but more is highly desirable because it directly improves the performance of the whole communications system. The gain has to be at least 25 dB, but several dBs more of gain is an advantage – they serve as a reserve against gain drop due to process variation or high temperature. The PA should be selective at the image frequency of 51 GHz, because the filter can't provide sufficient image rejection. Gain compression at 51 GHz, compared to 61 GHz should be at least 10 dB. Input and output matching is done for 50 Ω.

Other parameters such as saturated output power are not very important. The chip has only passive cooling. This limits maximum power consumption due to self heating effect.

5.4.2. PA Schematic Design

P1dB (or P_{sat}) is usually the most important PA parameter, and it is a vital parameter for the performance of the whole wireless system. The design of a PA at 60 GHz is centred around optimizing the last stage for maximum output power, P1dB (or P_{sat}), because at high frequencies HBT transistors have limited performance (lower breakdown voltage – lower voltage swing) limiting the output power. The design entails making several decisions regarding technology, PA topology, PA class, possible power combining. Then it can move on to the schematic optimization, layout drawing and post-layout simulations. This phase is usually iterative, because schematic simulations can't include the layout effects, and have to be redone when the layout is ready. Different optimizations may be necessary to improve PA performance regarding some parameter (output power, gain, input/output matching, PAE, layout size…).

The following decisions were made for the fabricated PA:

- **Technology choice.** The used technology is IHP's SiGe:C BiCMOS H1 technology. It is used for the whole circuitry, and this decision was made on the beginning of the WIGWAM project.

- **PA topology.** Knowing that the HBTs have limited performance regarding the output power, PA topology had to offer good performance regarding the output power. There are two obvious candidates regarding the choice for PA topology: common emitter and cascode. Common emitter allows operation under low supply voltage, but cascode allows much large voltage swing at the output, because collector-base breakdown voltage (for common base) is much larger than collector-emitter breakdown voltage (for common emitter). For IHP HBTs, BV_{CBO} is 4.5 V and BV_{CEO} is 1.9 V. To achieve the same output power as cascode, common emitter topology has to use transistors with more fingers or more transistors in parallel. However, the number of parallel transistors that can efficiently used is limited, as was discussed in the section on power combining techniques. This means that higher output power can be achieved with the cascode for the same efficiency. Another important point is that the PA has differential topology. Differential PA has up to 3 dB more power, which is combined using a differential antenna. Differential topology also has common-mode noise rejection and for a symmetrical layout inherent ac ground for the differential signal on the symmetry line.

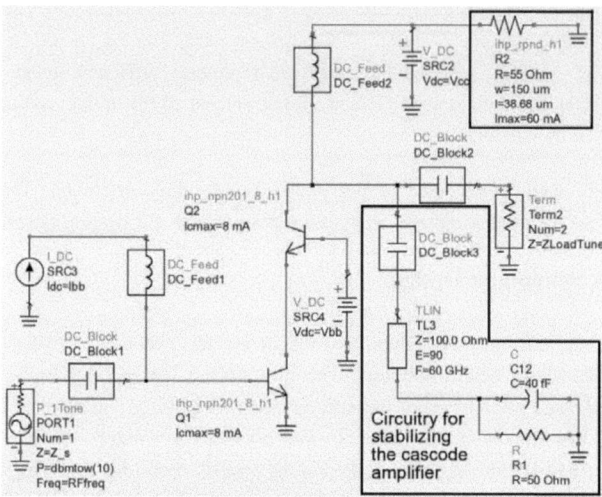

Figure 5.14 Simplified ADS schematic used to simulate the optimum output impedance for P1dB power level. Each transistor in the schematic represents four transistors with eight fingers. The encircled resistor R2 is used to model the double power consumption and related self heating in the differential circuit.

- **PA class.** As pointed out earlier, the application of interest in our case is amplification of an OFDM signal, which is sensitive to nonlinear distortion, meaning that the design will focus on optimizing the PA for the maximum P1dB. Class A was chosen because it offers the highest linear output power.

- **Power combining.** Simulations showed limited output power even with the largest, size 8, HBT transistor npn201_8. As discussed in the section on power combining, the number of parallel transistors or amplifiers that can be combined with reasonable efficiency is very limited. Apart from the limiting factors mentioned in the power combining chapter, one more is important in our case. We have a fully integrated transmitter, mounted on a board with only passive cooling, and self heating is an important factor here. Increasing the number of the transistors in the output stage leads to higher power dissipation and self heating. Rising temperature reduces both the gain and the output power of the PA. Simulations showed that a good compromise between maximizing the output power and keeping reasonable PAE is 4 parallel npn201_8 transistors in the last stage for each output, i.e. 8 transistors for the differential output.

The program used for the simulations is Agilent's ADS (Advanced Design System). The schematic used to test the performance of the last stage is shown in Figure 5. 14. At this design stage, biasing was ideal – with ideal current and voltage sources and dc block and feed elements. Each transistor in the schematic represents four parallel transistors (multiple factor of 4).

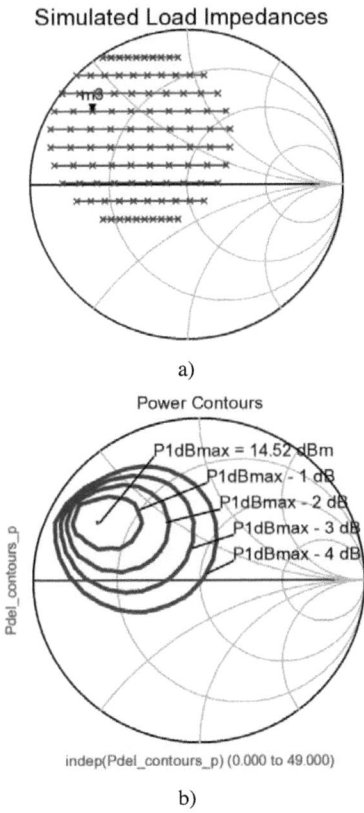

Figure 5. 15 a) The marked points in the Smith chart represent the values of the load impedance for which the P1dB was simulated. b) Power contours representing the load impedances for optimal P1dB (14.52 dBm) and 1,2,3 and 4 dB lower P1dB values.

S-parameter simulation showed that the cascode transistors are potentially unstable at lower frequencies. The circuitry added to stabilize the transistor is shown in the schematic. The circuitry has almost no effect on the S-parameters at 60 GHz. It was later removed, because the output matching structure stabilizes the cascode at low frequencies.

Figure 5. 15a shows the load impedances in the Smith chart that P1dB was simulated for. For each of the shown load impedances, optimum input source impedance was found and P1dB calculated. The results of these simulations are power contours in Figure 5. 15b.

After optimizing the quiescent points of the transistors, for the supply voltage of 4 V, the collector current of one transistor is 15 mA. The resulting maximum P1dB is 17.51 dBm, which corresponds to 20.51 dBm for the ideal differential case.

It is of interest to see how this maximum output power of 20.51 dBm is reduced as ideal assumptions are replaced with real ones in the simulation. Firstly, differential circuit has double power consumption, and hence, more self heating. To simplify the simulation, double power dissipation is introduced with an additional resistor with the same current and dissipation as the transistors (see Figure 5. 14). The differential P1dB is reduced in simulation to 20.05 dBm.

The next assumption is that the simulation temperature is not 25 °C, but 80 °C. The value of 80 °C is not calculated, but pessimistically estimated based on previous experience with chips of similar size, power consumption and boards used for mounting. The chip size and power dissipation referrer to the whole transmitter chip, rather than just the PA. With the environment temperature of 80 °C, the differential P1dB is reduced to 18.67 dBm.

The connections between the transistors are ideal. After putting the transistors in the layout, and drawing the connections, the connections are simulated with the EM simulator Momentum, which is part of the ADS. The transistors are then simulated with these connections, and the resulting differential P1dB is further reduced to 17.52 dBm (see Figure 5. 15b). The optimum output impedance is also slightly changed.

The P1dB has so far been reduced by 3 dB compared to the ideal case. Further reduction of the P1dB is expected in the real output matching network, after adding additional stages for more gain, and finally due to the imperfect antenna matching.

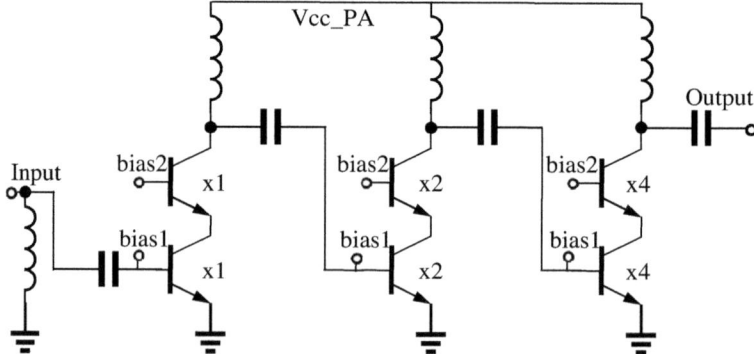

Figure 5. 16 Simplified PA schematic showing one half of the differential PA topology.

The amplification from one cascode stage of approximately 11 dB is too small, and three stages are required to achieve minimum 25 dB of gain. As discussed in the section on the linearity of cascaded amplifiers, cascading amplifier stages lowers the P1dB of the whole amplifier. Table 5. I shows that for three stages, gain of 10 dB per stage and P1dB difference between consecutive stages of 3 dB, total P1dB will be reduced by 0.66 dB. To achieve 3 dB P1dB difference between stages, second PA stage has half the number of parallel transistor (i.e. two), and the first stage half of that – i.e. one transistor. Half of the transistor number gives half of the current and with the same voltage swing results in half of the P1dB power, i.e. 3 dB less.

The three-stage PA topology is shown in Figure 5. 16. Only one half of the differential PA is shown – for simplicity. There are no current sources in the emitter branch, making the PA pseudo differential. The current sources are omitted to increase the voltage headroom and reduce the power consumption/self heating. The common mode rejection is not required from the PA – because very low common mode signal is expected at the input (ideally zero), and the on-board antenna is differential Vivaldi, with inherent common mode rejection. The PA layout is, however, optimized for the differential gain, and the PA exhibits limited common mode rejection in the measurements.

The next step is design of the output matching network, which will provide the simulated optimal load impedance for the transistors. After analyzing different matching structures, the L-C matching structure shown in Figure 5. 16 was chosen. This L-C structure is also used for inter-stage and input matching due to its advantageous characteristics. It is chosen because it performs different functions and has advantageous characteristics:

1. It performs the output, input and inter-stage matching.

2. Calculated values for the Ls and Cs are easy to realize. Inductors are around 60 pH and the input inductor 120 pH. These values are easy to realize, and have good Q factor. Capacitors have values around 100 fF – also easy to realize.

3. The matching structures can be easily realized as symmetric with ac ground on the symmetry line of the PA layout.

4. It feeds the dc current to the transistors, replacing the ideal dc feed element in Figure 5. 14.

5. It performs ac coupling at the output, input and inter-stage, replacing the ideal dc block elements in Figure 5. 14.

6. The inductors in the collector branch have small inductance, and are for lower frequencies effectively short. The simulation shows that they stabilize the cascode for lower frequencies, replacing the stabilizing circuitry in Figure 5. 14. The stabilizing circuitry is hence removed.

7. The input inductor is shortened to the ground, and as such is an ideal ESD input protection. This is, however, not important for the integrated PA.

The inter-stage matching is optimized for the PA frequency characteristic. It should be flat around 60 GHz, but it is desirable to have suppression of the image – at 51 GHz. The goal is at least 10 dB suppression of the image signal. The L-C matching structure is selective enough to achieve this goal easily.

The PA has two supply voltages Vctrl_PA and Vcc_PA. Vctrl_PA controls the dc current of the transistors, by controlling the biasing of the common emitter transistors (bias1 – see Figure 5. 16). It draws a small current of 3 mA. Vcc_PA is connected to the collector branches of the common base transistors and to their biasing circuitry (bias2 – see Figure 5. 16). Vcc_PA draws most of the current – approximately 200 mA. The PA is designed for the equal value for Vctrl_PA and Vcc_PA of 4 V, so that they could be connected. The voltages were separated so that the supply voltage (Vcc_PA) and the dc collector current can be set separately for measurements tests.

5.4.3. PA Layout and Post-layout Simulation

Drawing the layout for a high frequency circuit is often not easy, and involves different challenges. The goal is to create the layout that corresponds to the schematic. The layout components, however, suffer from unwanted parasitics (capacitance or inductance) and mutual coupling, which are not present in the schematic. Parasitics are especially important at high frequencies such as 60 GHz where, for example, a small capacitance of 5 fF has impedance of just 500 Ω. Such capacitance in an oscillator core can cause large oscillation frequency shift.

Another common problem is that certain schematic elements are either too big or too small when realized in the layout. In these cases, one has to often re-simulate the schematic and try to find structures that are easier for layout realization. The big structures require not just more area, but often need to be placed further away from the circuit core and connected with long lines, which then have to be modelled. As for the small elements, they are very sensitive to parasitics and process variations. If a capacitor is very small, it can be exchanged with two double size capacitors placed in series. This will make the total capacitance more robust to parasitics and process variations.

Due to these problems, making the layout is usually an iterative process. The schematic may have to be changed to facilitate making the layout, or a necessary long connection in the layout has to be introduced in the schematic. Both require schematic re-simulation.

The layout should be simulated (EM simulation), to include all parasitics and couplings. It is then imported in the schematic (as an *n*-port data file), and the whole circuit can be simulated. The layout is often changed and re-simulated to optimize one or more circuit parameters.

For the PA presented in this chapter, the schematic had to be changed, because the stages were too close causing strong inter-stage coupling and concentrating power dissipation in a small area. To alleviate these problems, the stages were separated by 200 μm and connected with TLs. The TLs were introduced in the schematic and it was re-optimized.

The layout was made in stages. The layout is drawn and the output matching structure, together with the output pad is simulated in the Momentum. The simulated structure is then simulated in the schematic together with the lumped elements (capacitors). The resulting S-parameters are compared with the S-parameters of the output matching structure from the schematic. If the matching is not good, the layout is optimized and re-simulated. Figure 5.17 shows the comparison of S-parameters from the output matching structure in the schematic and from the layout, which includes the pad and transmission line to the pad. S_{11} shows the impedance seen by the transistor, and S_{22} the impedance at the output pad. This kind of procedure is done for both inter-stage matching structures and the input matching structure.

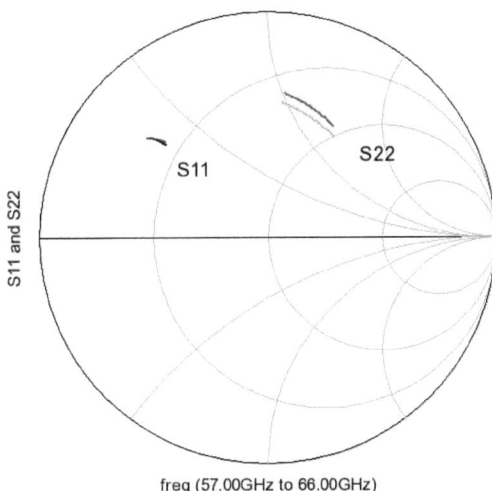

freq (57.00GHz to 66.00GHz)

Figure 5.17 Comparison of the simulated S-parameters of the output matching network with the output transmission line and the RF pad in the schematic (lumped elements) and layout (distributed elements).

Another very important aspect of drawing the PA layout is assuring that the PA has good ground connection. This is not so problematic for the on-wafer measurements of the PA, but it is critical when the PA is integrated in the transmitter. The transmitter will be mounted on a board and bonded. Bondwires have minimal length of 500 μm, which corresponds to inductance of 500 pH. At 60 GHz this is impedance of 188 Ω. This is very large impedance and the circuit would not work with such ground connection. The inductance would also most likely make the PA unstable.

Two measures are taken to improve the ground connection. One is related to the TX layout, and it involves placing many ground pads in the layout. Having many ground bondwires reduces the impedance of the ground connection. The second measure is to create a symmetrical PA layout, such that the symmetry line plays the role of the ac ground for the differential signal. If we take a look at the single-ended PA schematic in Figure 5.16, we see that good ground is needed for the RF signal at the emitters (of the common emitter topology), at the inductors in the collector branches (Vcc_PA is ground for the ac signal) and at the bases of the common base transistors. This means that these points need to be close to the symmetry line in the layout, so that the current of the differential signal flows from one half of the PA into the other. In the ideal case there should be no RF current flowing into the ground, creating no ac voltage drop on the bondwires.

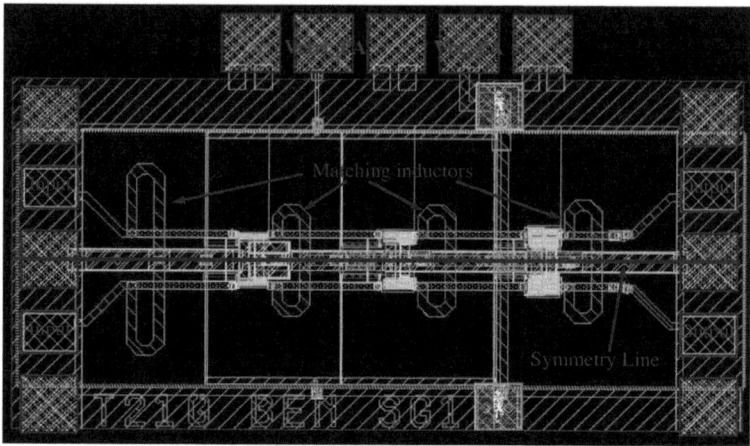

Figure 5. 18 Power amplifier layout

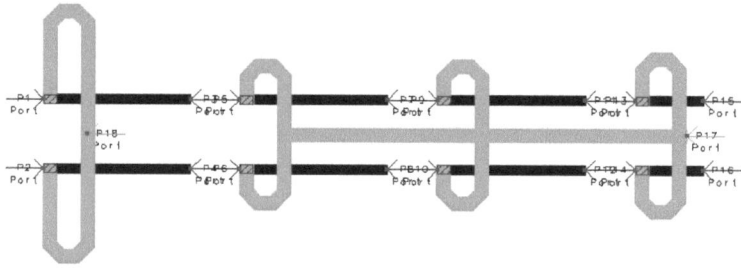

Figure 5. 19 Structure for ADS Momentum EM simulation. Green is top metal layer; blue is one metal layer lower.

PA layout is shown in Figure 5. 18. The red line shows the symmetry line in the layout. The inductors from the matching structures are realized as bent lines, so that they end at the symmetry line. The transistors are also placed close to the symmetry line, so that the emitters and bases have short connections.

The PA layout was in the end simulated to check the performance and to make sure that the whole PA is stable. Figure 5. 19 shows the structure that was simulated in Momentum. This is a simplified layout, because simulating the whole PA layout with all connections and vias would require much larger computer resources than were available, and not all connections are required for the analyses. Part of the stability check is to simulate whether there is any possibility of oscillations on the power supply lines. This is done by performing the S-parameter simulation when the ports are connected to the supply lines (Vctrl_PA and Vcc_PA) and the input and output are terminated with 50 Ω [5.3]. This simulation also showed unconditional stability of the PA.

5.5. Power Amplifier Measurement

The PA was first measured on-wafer. After that it was integrated in the transmitter and the transmitter was mounted on the Rogers board and bonded. The output power of the transmitter board was measured, and by doing so the PA output power on-board performance was measured. The temperature measurement of the transmitter board with an IC camera was also done. Finally the PA is compared with the state-of-the-art of mm-wave PAs fabricated in SiGe and CMOS technologies.

The chip photo of the fabricated PA is shown in Figure 5. 20. Chip dimensions are 1×0.58 mm^2 = 0.58 mm^2, with pads, and 0.7×0.3 mm^2 = 0.21 mm^2 without pads.

5.5.1. Measurement Setup

The stand-alone PA was measured only on-wafer. Two measurement setups were used. Figure 5. 21 shows the setup for the measurement of S-parameters. It characterized the PA for the small signal operation. The Agilent 8510XF vector network analyzer offers single-ended measurements up to 110 GHz. The setup uses Picoprobe RF probes with 1mm coaxial connector for the input and output signals.

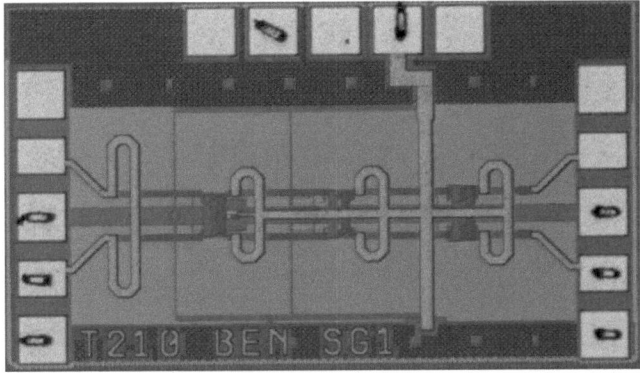

Figure 5. 20 Power amplifier chip photo.

Figure 5. 22 shows the measurement setup for the measurement of the saturated output power and P1dB. Agilent E8257D PSG analog signal generator (up to 70 GHz) is used for the input signal generation. The setup uses Picoprobe RF probes with V-band (50-to-75 GHz) waveguide connector for the input and output signals. The output probe is connected to the Agilent V-band V8486A power sensor with waveguide input. The sensor is further connected with the Agilent E4419B power meter. Measuring output power level with the power sensor and power meter is the most accurate approach. The measured value needs to be corrected by only the value of the output probe insertion loss. Output power could also be measured with the network analyzer, but the input power level is often not sufficient, the network analyzer input can be damaged by the strong output signal, and the accuracy is lower compared with the power sensor method.

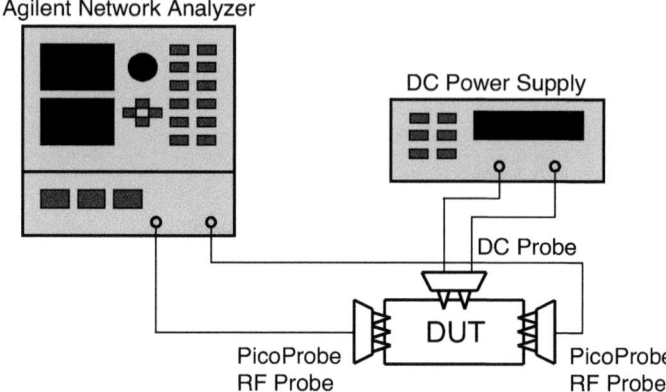

Figure 5. 21 Small-signal S-parameter measurement setup for on-wafer measurement. The DUT is the power amplifier.

The PA output spectrum is also checked for any oscillations with the R&S FSEM30 spectrum analyzer. The analyzer uses special external mixer for the V-band range with the waveguide input.

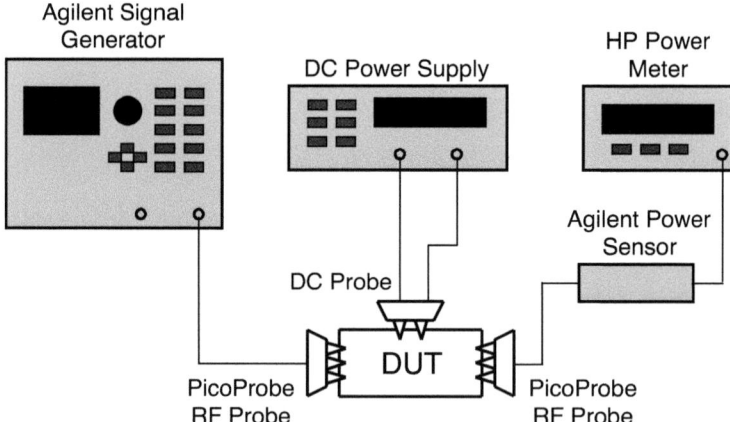

Figure 5. 22 Large-signal measurement setup for on-wafer measurement. The DUT is the power amplifier.

5.5.2. Measurement Results

The PA was measured for the supply voltage of 4 V and dc current of 200 mA – 800 mW power consumption; and for the supply voltage of 3.7 V and dc current of 162 mA – 600 mW. Measured and simulated S-parameters for 800 mW power consumption are shown in Figure 5. 23 and Figure 5. 24.

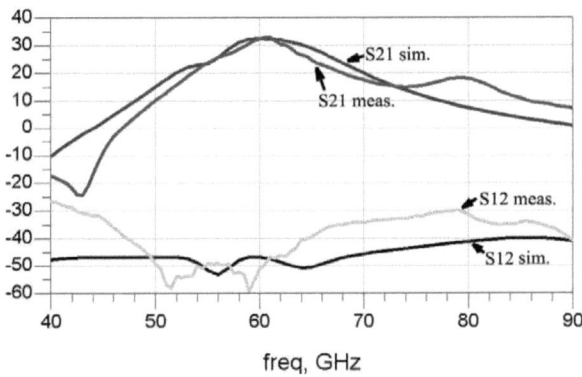

Figure 5. 23 Comparison of measured and simulated S21 and S12 parameters.

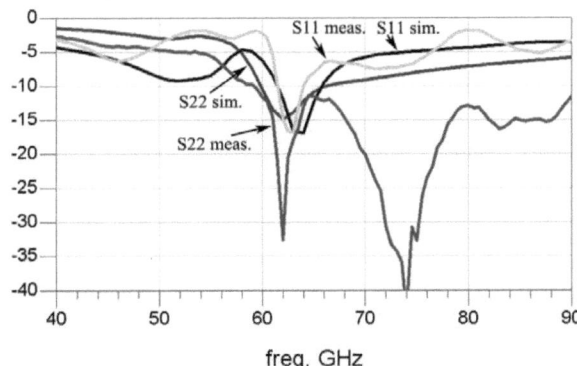

Figure 5. 24 Comparison of measured and simulated S11 and S22 parameters.

The PA has high gain of 33 dB at 61 GHz It exhibits image rejection at 51 GHz with respect to the signal at 61 GHz of 20 dB (Figure 5. 23). Isolation is very good (better than –40 dB up to 65 GHz). These results are in good agreement with simulation results, which confirms that the small signal models and EM simulation are accurate enough. Small signal S-parameters do not change much for smaller power consumption (600 mW) – amplification peak is approximately 2 dB lower.

S-parameters were also measured while probing the input pad of the one half of the differential circuit and the output pad of the other half of the differential circuit. The measured gain at 61 GHz is 30 dB – i.e. 3 dB lower than when the corresponding output is probed. Assuming that probing one input corresponds to probing both inputs lets us calculate the differential and common mode amplification and CMRR. If one half of the input signal is the differential signal and the other half the common mode signal, at one input they are in phase (adding power) and at the other input out of phase (cancelling each other). The measured amplifications are then in one case the sum and in the other case the subtraction of the differential and common mode amplifications. The calculation shows that the differential gain is 34.7 dB, common mode gain is 19.3 dB and CMRR is 15.4 dB. These values should be taken only as approximate ones, because the assumption under which they were calculated does not correspond to the real case.

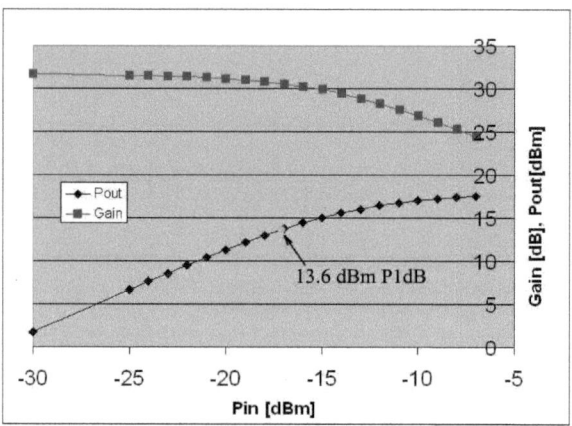

Figure 5. 25 PA output power and gain vs. input power at 61.5 GHz for 800 mW power consumption.

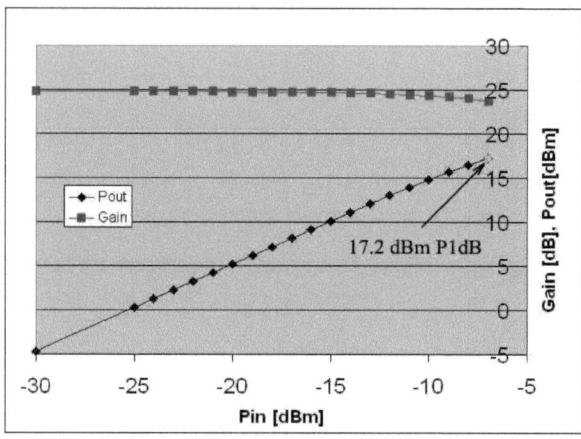

Figure 5. 26 PA output power and gain vs. input power at 65 GHz for 800 mW power consumption.

The Agilent network analyser can generate input signal up to –7 dBm, which in combination with strong PA amplification allows measurement of the PA P1dB with this setup (Figure 5. 21). Input power sweep measurements at 61.5 GHz and 65 GHz are show in Figure 5. 25 and Figure 5. 26. PAE measured for the input power sweep at 61.5 GHz for power consumption of 600 mW is shown in Figure 5. 27. Summary of measured results in the frequency range from 59 to 66 GHz for 800 and 600 mW power consumption are shown in Table 5. II and Table 5. III, respectively. The values for the saturated output power shown in these tables are measured using the power sensor (Figure 5. 22). Figure 5. 28 shows the measured AM-to-PM distortion of the PA at 61.5 GHz. The plot represents the relative shift of the phase of the S_{21} parameter measured for the input power sweep.

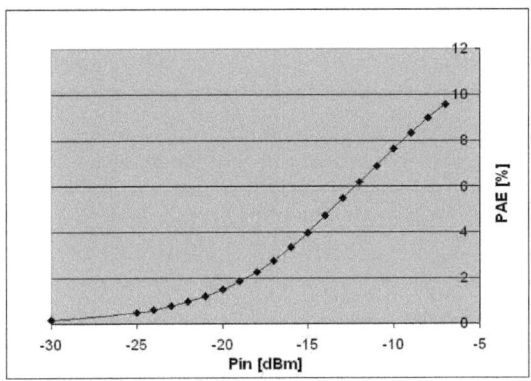

Figure 5. 27 PA power-added efficiency at 61.5 GHz for 600 mW power consumption. Maximum PAE of 10.2 % is reached for input power of –3 dBm.

Freq. [GHz]	Max Gain [dB]	P1dB [dBm]	Psat [dBm]	PAE [%]
59	30.7	12	17	6.3
60	32.4	11.7	17.4	6.9
61	32.9	12.8	17.8	7.5
61.5	31.7	13.6	17.9	7.8
62	30.5	14	18.2	8.2
63	28.9	15.2	18.5	8.9
64	26.4	16.1	18.7	9.4
65	24.8	17.2	18.9	9.8
66	22.8	17.1	19.1	10.2

Table 5. II Measured PA performance for 800 mW power consumption.

Tables 5. I and II show large change in P1dB with the frequency. This is the result of the single-ended measurement of the differential circuit, which effects the P1dB measurement. The measurement of the saturated output power of the whole TX chip (effectively of the integrated PA) is shown in Figure 5. 29. The integrated PA has differential input and output (differential on-board Vivaldi antenna). The measurement shows that the PA gives the highest differential output power of app. 20 dBm at 59 GHz. The P1dB is approximately 3 dB lower, i.e. 17 dBm. There are two main reasons for the different results of the on-wafer measurement and the on-board measurement. One is that the on-wafer measurement is single-ended and the on-board measurement is differential. The other is that the antenna matching is narrow band – simulated for 60 GHz. We should also keep in mind that the integrated PA works at higher temperature compared with the on-wafer PA (see temperature measurement of the TX board in Figure 5. 30), that the ground connection is not as good as for the on-wafer measurement and that the antenna matching doesn't provide exactly 50 Ω matching.

Freq. [GHz]	Max Gain [dB]	P1dB [dBm]	Psat [dBm]	PAE [%]
59	29.9	12.3	17	8.5
60	31	12	17.4	9.2
61	30.7	13.2	17.7	10
61.5	29.4	13.5	17.9	10.2
62	28.3	14	18	10.6
63	26.8	14.3	18.3	11.2
64	24.7	14.6	18.3	11.3
65	23.2	15.3	18.4	11.4
66	21.4	15.5	18.4	11.5

Table 5. III Measured PA performance for 600 mW power consumption.

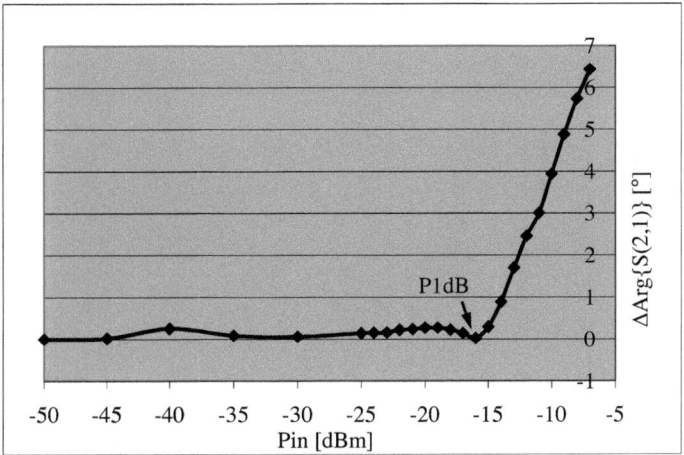

Figure 5. 28 Measured AM to PM distortion of the PA.

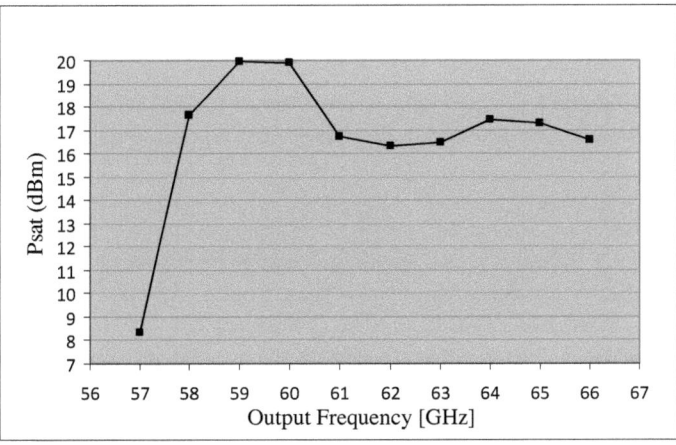

Figure 5. 29 Saturated output power at the output of the TX (PA) versus IF input frequency from 1 to 10 GHz. The PLL up-conversion frequency is 56 GHz resulting in the output frequency range of 57 to 66 GHz.

Figure 5.30 Temperature measurement on the TX board done with an IC camera showing chip temperature of 60,6°C.

5.5.3. Comparison with the State-of-the-Art

A large license free ISM band of 7 GHz, centered around 60 GHz, offering an opportunity for wireless communication with data rate of more than 1 Gbit/s, has been the driving force for the development of the 60 GHz communication systems. Advances in SiGe Bipolar/BiCMOS technologies, with hetero-junction bipolar transistor (HBT) cut-off frequencies as high as f_T/f_{max} = 300/350 GHz, made them competitive with III/V technologies for this frequency range. The increase in f_T/f_{max} of HBTs has a consequence in reduction of breakdown voltages (BV_{CEO} and BV_{CBO}). This and relatively low amplification at these frequencies, make achieving high output power – the main purpose of a PA – challenging. Fully integrated 60 GHz chipsets in BiCMOS technologies for wireless communication with data rates in the order of 1 Gbit/s have already been reported [5.4], [5.5].

A number of papers on PA in SiGe design for 60 and 77 GHz (automotive radar) has been presented [5.6] – [5.13], with P1dB as high as 14.5 dBm [5.10] and maximum saturated output power of 20 dBm [5.6]. PAs in CMOS have also been presented [5.14] – [5.16], but with considerably lower performance compared with SiGe PAs (P1dB of 9 dBm and Psat 12.3 dBm). Table 5.IV presents the comparison of the mm-wave PAs, which were presented before this PA was published. This PA has the highest reported P1dB, at the time it was published. It also has the highest gain, but the peak PAE is lower due to class A operation and power hungry first two stages. The on-board TX measurements have also shown saturated power of 20 dBm and P1dB of 17 dBm at 60 GHz.

We should also notice that the work on achieving higher output power has since been focused on using beam forming and power combining in the air, rather than using larger number of transistors and amplifiers for the on-chip power combining.

Freq. [GHz]	Technology [μm]	f_T/f_{max} [GHz]	Mode of operation	Max Gain [dB]	P1dB [dBm]	Psat [dBm]	Peak PAE [%]	Reference
61.5	SiGe 0.25	200/200	differential	31.7	13.6	17.9	7.5	This PA [5.17]
65	SiGe 0.25	200/200	differential	24.8	17.2	18.9	9.8	This PA [5.17]
60	SiGe 0.13	200/240	differential	18	13.1	20	12.7	[5.6]
60	SiGe 0.18	130/120	balanced	12	11.2	15.8	16.8	[5.7]
61.5	SiGe 0.13	285/207	differential	12	8.5	14	4.2	[5.8]
77	SiGe 0.13	175/265	differential	-	-	18.5	5.4	[5.9]
77	SiGe 0.13	285/207	single-ended	17	14.5	17.5	12.8	[5.10]
77	SiGe 0.13	200/240	differential	6.1	11.6	12.5	2.5	[5.11]
58	SiGe 0.13	200/240	single-ended	4.2	-	11.5	20.9	[5.12]
85	SiGe 0.13	285/207	single-ended	8	-	21	3.4	[5.13]
60	CMOS 0.09	-	diff.-to-single	5.5	9	12.3	8.8	[5.14]
60	CMOS 0.09	-	single-ended	8.3	8.2	10.6	5	[5.15]
52	CMOS 0.09	180/107	single-ended	25	5	8	7	[5.16]

Table 5. IV Comparison of mm-wave power amplifiers.

5.6. Summary

This chapter presents the design of a 60 GHz PA in the IHP 0.25 μm SiGe:C BiCMOS process with f_T/f_{max} = 200/200 GHz. The theory of a PA design was addressed with an emphasis on the aspects that are the most important for a mm-wave PA design. Stability issues and nonlinearities of a single stage and cascaded PA have been presented.

Two main power combining techniques of paralleling transistors or amplifiers have been presented. Both have limited power combining capacity for high frequencies and integrated solutions. For higher number of parallel transistors or amplifiers the efficiency drops and self heating reduces output power. Self heating is a limiting factor for an integrated PA with only passive cooling.

Modeling of passive structures was presented, as well as the test structures used to verify EM simulations and measurement procedure.

The procedure of the PA design was presented. The differential cascode PA topology was chosen with class A operation and power combining of paralleling transistors was implemented. PA simulations for the maximum P1dB were presented and the effects that degrade it by 3 dB were analyzed.

Difficulties related to crating a mm-wave layout from schematic were discussed. The layout is drawn and optimized in an iterative design procedure and post layout simulation was done. The PA features a symmetrical layout, which inherently gives ideal ac ground on the symmetry axis. This is very important for an wire-bonded chip working on high frequencies, where ground connection via bond wires is poor.

The measured PA had the highest reported P1dB of 17 dBm when it was published. Saturated power is 18.7 dBm at 61.5 GHz and maximum 19 dBm at 66 GHz. Measured amplification is 33 dB at 61 GHz. On board TX measurements of have shown saturated power of 20 dBm and P1dB of 17 dBm at 60 GHz. PA image rejection is as high as 20 dB at 51 GHz which is 10 dB better than required. Compact differential design resulted in small size of 1 mm×0.6 mm=0.6 mm^2 with pads. Maximum PAE is 7.8% for 800 mW power consumption, and 10.2% for 600 mW.

Chapter 6

Transmitter Integration

6.1. Introduction

In this chapter the integration of TX chips is presented and the most important aspects related to the integration and board design are discussed. All three TX versions are presented with their measurement results. The simulated and measured link budget are compared. The achieved results are compared with the state-of-the-art.

6.2. TX Integration and Board Design

Before integration, TX components have to be simulated together. Different simulations regarding the frequency characteristic and output power are performed for each TX version. The PLL is, however, not simulated together with the rest of the TX, because the time domain simulation for the PLL settling would take too much time. The PLL settling is simulated separately, and in the TX simulations the PLL is replaced by its VCO or ideal signal source. It should be said here, that the TX simulations serve to check that all components work well together. Their performance parameters have earlier been calculated, and there shouldn't be any need for their optimization in this design stage.The material used for the board is Rogers3003. It has low loss and low electric permeability at 60 GHz. Since the thickness is just 125 μm, another layer of FR4 was used for mechanical stability. FR4 thickness is 1200 μm. Both Rogers and FR4 have top and bottom metal (copper) layer. These two boards are connected with 100 μm thick layer of Prepreg.

As mentioned in the PA chapter, the TX and especially the PA require good on-chip ground for good operation. To achieve this the ground connection needs many short bondwires. The TX chip has as many ground pads as possible. To reduce the bondwire length, the chip is placed in a cavity. The chip is thinned (to 250 μm) so that the pads are in the same level as the top metal on the board. The cavity is encircled with a ground ring to achieve minimum bondwire length. Figure 6. 1 shows a close-up photo of a bonded TX chip. The cavity reaches the top FR4 metal layer, which is a ground layer. Cavity walls are metallized, so that the ground ring has good and broad ground connection to a ground plane.

Top FR4 metal layer is also used to conduct heat from the chip. It is connected with the bottom metal plate (bottom FR4) with many vias so that the heat can be radiated on the other side of the board. Figure 5. 30 is a thermal photo of a TX board, showing TX chip temperature and how the heat is conducted away from the chip.

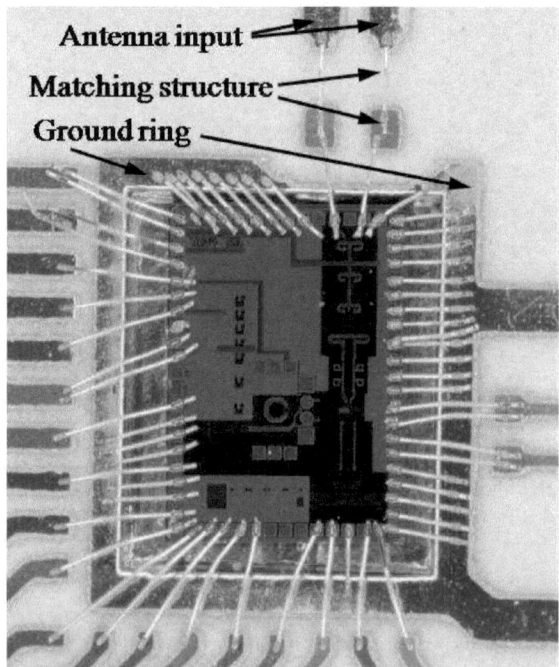

Figure 6. 1 Close–up photo of a bonded TX chip.

6.3. Measurement Results

TX chips were measured only on-board. The chips are big, and the lack of appropriate probes made on-wafer measurement impossible. The chips were first tested with a sinus signal. The boards were then measured together with RX boards for maximum data rate and distance.

6.3.1. TX Version I Measurement Results

Figure 6. 2 shows TX version I chip photo with marked TX components. Chip size is $1.8 \times 1.6 = 2.88$ mm^2. The transmitter consumes 1150 mW from 3.7, 3.5, 3, 2.6 and 2.5 V supplies. Figure 6. 3 shows an assembled TX board with single-ended Vivaldi antenna.

The chip was first measured with a sinus signal. The chip worked well, but there was a problem with the PLL locking when the PA was on. The supply voltage of the dividers had to be fine-tuned in order to make the PLL lock. This problem was solved in the next version by separating PLL ground from the rest of the chip.

The influence of ground connection was also tested. When only a few long bondwires are bonded, output power of the chip is much reduced – around 10 dB. This effect was also present for the TX version II, but the power drop was smaller – around 4 dB. This is because the TX version II is fully differential, and the PA has virtual ground on the symmetry line.

Figure 3. X shows the measured phase noise. The measured image-rejection is better than – 40 dBc.

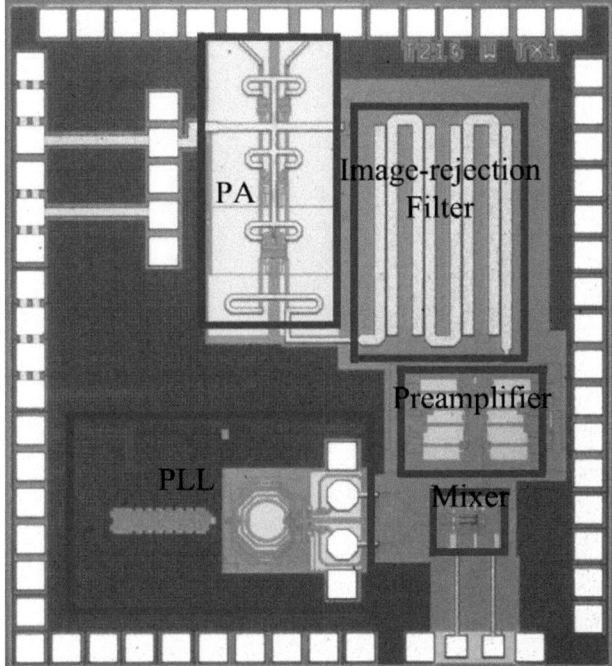

Figure 6. 2 Version I transmitter chip photo.

Block diagram of the measurement setup is shown in Figure 6. 4. A photo of the measurement setup in a laboratory is in Figure 6. 5.

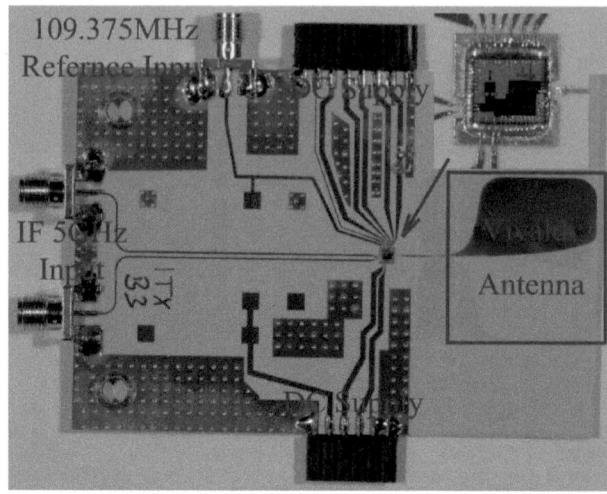

Figure 6. 3 Transmitter board photo with chip close-up.

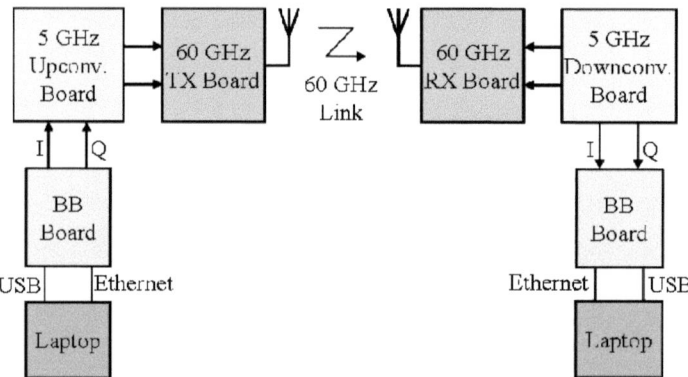

Figure 6. 4 Analog front-end measurement setup.

Data transfer with 360 Mbit/s over 5 meter distance is achieved. OFDM signal with QPSK modulation scheme and ¾ coding was used. Figure 6. 6 shows the measured constellation diagram. This TX was presented in [6.1].

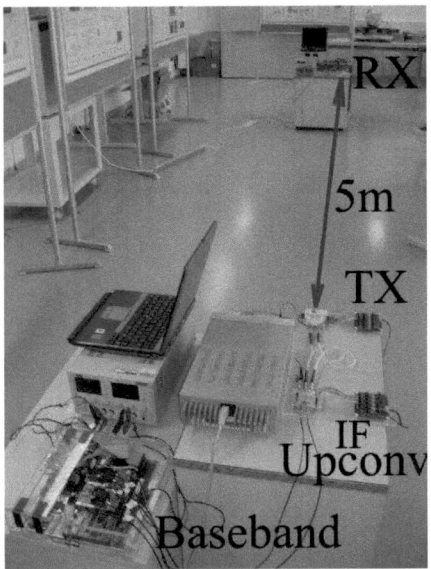

Figure 6.5 Version I AFE measurement setup photo.

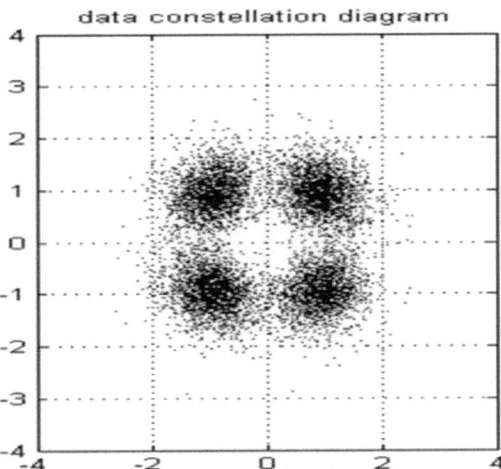

Figure 6.6 Constellation diagram for QPSK, ¾ coding, OFDM signal with 360 Mbit/s data rate over 5 m distance.

6.3.2. TX Version II Measurement Results

Figure 6.7 shows TX version II chip photo with marked TX components. The chip is larger than version I, because it has two filters. The size is $2.5 \times 1.6 = 4$ mm^2. It consumes the same power: 1150 mW.

The tests with a sinus signal showed proper functioning of the chip. Phase noise is the same as for the previous version. Image-rejection is below –40 dBc. Figure 5. 29 shows measured saturated output power for input frequency from 1 to 10 GHz. P1dB at 60 GHz is 17 dBm. Chip's temperature was measured with an IC camera (see Figure 5.30). The measured temperature is 60.6°C.

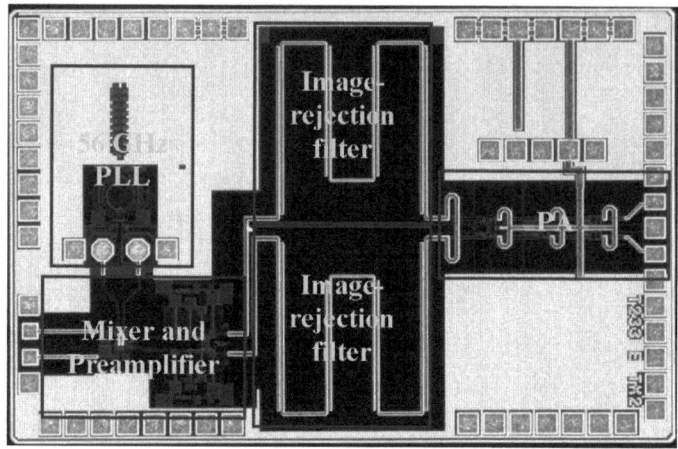

Figure 6. 7 Fully differential version I transmitter chip photo.

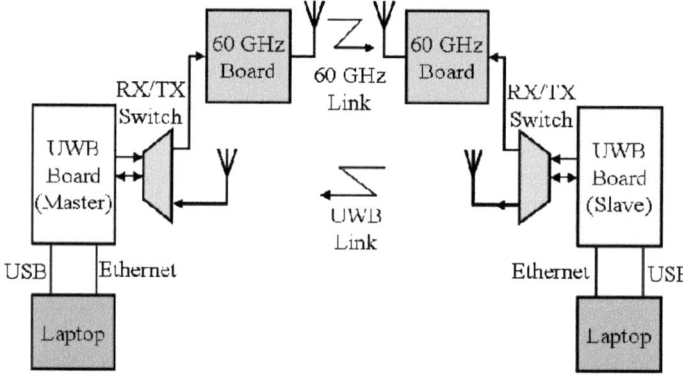

Figure 6. 8 Block diagram of the asymmetric dual-band demonstrator.

The TX was measured with a commercially available UWB system. The setup is presented at Figure 6. 8. It is an asymmetric demonstrator with a 60 GHz link in one direction, and a UWB link in the other direction. The UWB system has three channels – at 3, 4 and 5 GHz. Measured results for minimum and maximum distance for different data rates are presented in Figure 6. 9. The system was successfully tested for video transmission (video data rate 10 Mbit/s) up to 15 meters. The system worked better when both links were at 60 GHz, and video transmission was possible up to 60 meters. This was the case because the automatic gain control (AGC) in the UWB boards was designed for equal channel in both directions. The system was presented in [6.2].

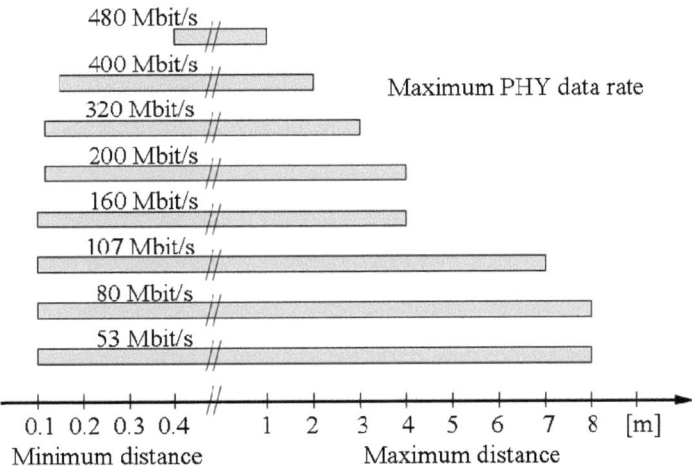

Figure 6. 9 Maximum and minimum distance of communication for different data rates.

6.3.3. TX Version III Measurement Results

Figure 6. 10 shows TX version III chip photo with marked TX components. Although the chip contains the IF circuitry, it is smaller than version II, because the filters for this IF are much smaller. The TX size is 2.1×1.5 = 3.15 mm^2. It consumes 1300 mW from three different supplies: 3.7, 3.3 and 2.5 V. This chip has SPI control of I and Q mixer gain and the sideband suppression.

Sinus tests shows proper functionality. The measured phase noise at 1 MHz offset is –94 dBc/Hz. The image is below the noise level of the spectrum analyzer, which means that the image-rejection is better than –50 dBc. The output power and P1dB are on the level of the previous TX version, because the PA and the antenna are the same.

The TX was measured in a loop: Matlab – Tektronix AWG generator – TX – RX – Agilent oscilloscope – Matlab. Matlab gives OFDM frames information to the AWG, which generates the frames and feeds them to the TX. The received signal is fed from the RX to the oscilloscope, and the sampled signal is analyzed by Matlab. Data transmission of 3.6 Gbit/s (4.8 Gbit/s raw – i.e. with coding) was demonstrated over 15 meters with zero FER. The FER was measured for 2000 frames. The OFDM signal used 16QAM modulation scheme with ¾ coding. SNR measured over 15 m was 10.2 dB, over 10 m 12 dB and over 5 m 13.8 dB. Measured constellation diagrams are shown in Figures 6. 13, 14 and 15. A paper describing this TX version will be submitted to SiRF conference.

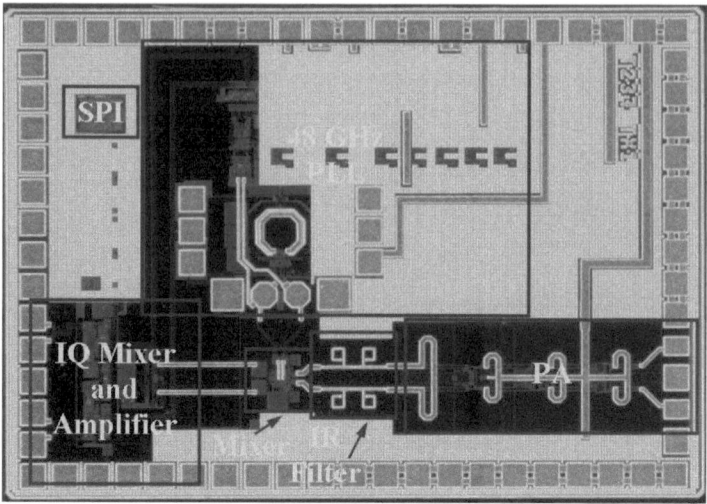

Figure 6.10 Version III fully integrated transmitter chip photo.

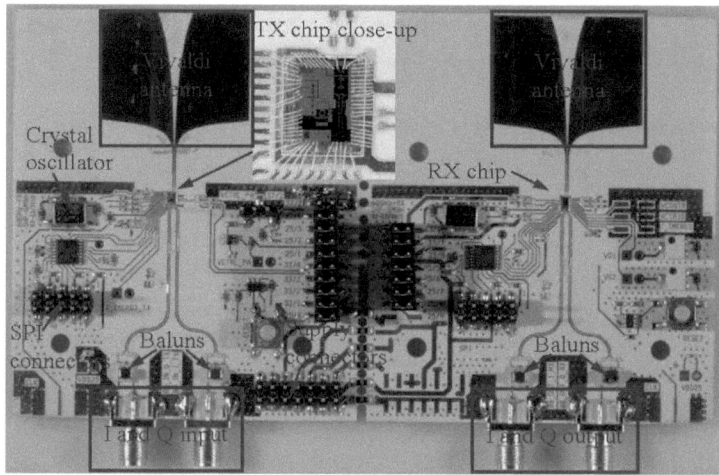

Figure 6.11 Fully mounted single-chip TX and RX board.

Figure 6. 12 Version III AFE measurement setup photo.

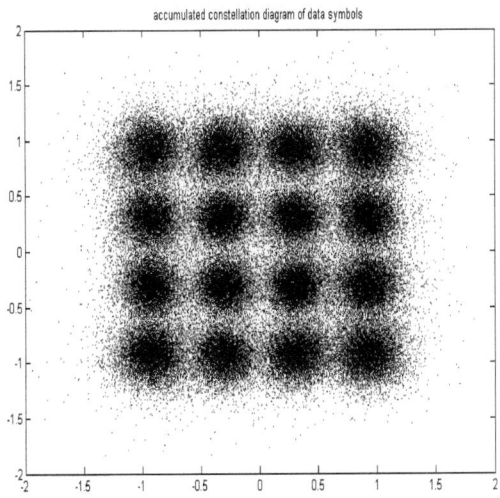

Figure 6. 13 Constellation diagram for 16QAM, ¾ coding, OFDM signal with 3.6 Gbit/s data rate over 5 m distance.

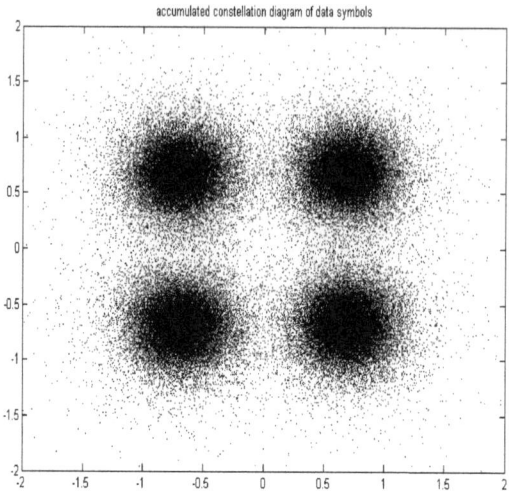

Figure 6. 14 Constellation diagram for QPSK, 2/3 coding, OFDM signal with 1.6 Gbit/s data rate over 15 m distance.

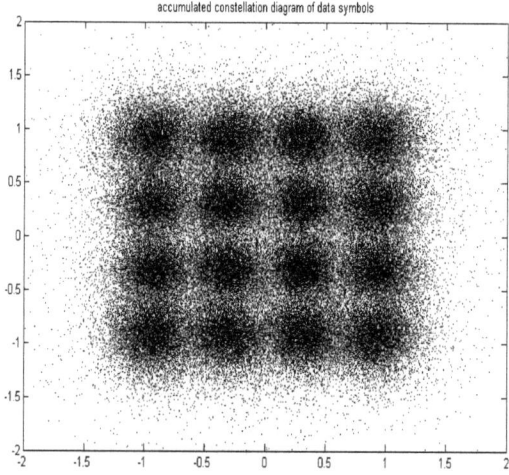

Figure 6. 15 Constellation diagram for 16QAM, ¾ coding, OFDM signal with 3.6 Gbit/s data rate over 15 m distance.

6.3.4. Link-Budget Calculation for the Version III AFE

The measurement results for the version III AFE are used here to calculate the link budget and to compare it with simulation. We will use measurement results for 5 and 15 m distance with 16QAM modulation scheme and ¾ coding. Both measurements were done with the same input level for the TX. This means that the SNR of the transmitted signal is the same for both measurements. Measured SNR over 15 m distance is 10.2 dB and from Equ. (2.4) we have:

$$10.2[dB] = 10\log(P_{RXin}/(P_{N,TX} + kTBF))$$ (6.1)

Measured SNR over 5 m distance is 13.8 dB. Since the distance is 3 times smaller, $P_{RX,in}$ and $P_{N,TX}$ are 9 times higher, which gives:

$$13.8[dB] = 10\log(9P_{RXin}/(9P_{N,TX} + kTBF))$$ (6.2)

RX noise figure is 8 dB, or linear 6.3. Bandwidth is 1.7 GHz, and RX temperature is 45 °C (T = 318 K). After solving the system with two unknowns ($P_{RX,in}$ and $P_{N,TX}$) we get that the signal power at RX input is $P_{RX,in}$ = –61.1 dBm, and SNR at TX output is 14.5 dB. The TX output power $P_{TX,out}$ is than from Equ. (2.1):

$$P_{TX,out}[dBm] = P_{RX,in} + L_{TX} - G_{ATX} + L_{FS}(d) - G_{ARX} + L_{RX}$$ (6.3)

TX and RX antennas are the same. Measured antenna gain at 60 GHz is 11.5 dBi. Loss of the RX and TX matching is estimated at 1 dB (somewhat more than ideal simulated case). Free-space loss over 5 m is 91.5 dB. TX output power is than calculated to be 9.4 dBm. The measured TX P1dB is 17 dBm, and simulated output back–off for 16QAM modulation should be 5 dB giving 12 dBm output power. The difference of 2.6 dB represents the implementation loss. It is relatively small, and it can said that the measured and simulated link–budget match well.

We should keep in mind that all the measured values (TX P1dB, RX noise figure, antenna gain) have certain margin of error. Exact values of TX P1dB, RX noise figure change from chip to chip and antenna gain from antenna to antenna. Matching is also different for each board, because it uses bondwires, and they differ due to manual bonding. Furthermore, the free–space loss doesn't take multipath into account, which can be both constructive and destructive. Consequently we can say that the difference of 2.6 dB between simulation and measurement is within the margin of error.

6.4. Comparison with the State-of-the-Art

The first 60 GHz integrated TX and RX chips were published in 2006, by IBM [1.10] and IHP [1.11]. The progress since then has been continuous, and three main trends can be observed:

1. Increase in the performance, i.e. larger data rates and larger communication distance.

2. Shift from SiGe to CMOS technology. The first published TX and RX chips were in SiGe, but the majority of the lately published 60 GHz chips are produced in CMOS.

3. Increase in the level of integration. Newer versions usually integrate IF circuitry and some even ADC and DAC.

4. TX and RX feature beam-forming topology. Beam-forming is implemented to boost the link-budget and to allow NLOS communication.

Table 6. I shows a comparison of the new version of AFE with version III TX with the state-of-the-art. Four reported systems have one TX and one RX antenna [6.3] – [6.6], and two systems implement beam-forming [6.6], [6.7].

Beam-forming chips are much larger and consume much more power than the non beam-forming ones. As such, they belong to a different class and shouldn't be directly compared. The chipset in [6.7] has 16 elements (TX and RX each), but consumes 6.2W and has an area of 43.8 mm^2. Detailed data rate measurements have not yet been published, except that the system can transmit 5.6 Gbit/s over an unspecified distance.

The results that are presented here will be published in a paper submitted to SiRF 2011 conference.

Reference	Technology [μm]	Topology	Power Dissipation [mW]	Chip Area [mm^2]	Data Rate [Gbit/s]	Max Distance [m]
This work	SiGe 0.25	Sliding IF	TX: 1300	TX: 3.15	3.6	15
[6.3]	SiGe 0.13	Sliding IF	TX: 822 RX: 547	TX 6.4 RX: 5.6	2	3.5
[6.4]	CMOS 65 nm	ZIF	TX&RX 374	TX and RX: 1.04	3.5	2
[6.5]	CMOS 90 nm	ZIF	TX&RX 300	TX and RX: 6.88	4	1
[6.6]	CMOS 90 nm	ZIF	TX&RX 200	TX and RX: 6.25	3.5 (raw)	Cable connection
[6.6]	CMOS 90 nm	Beamforming 4 Elemnts	N.A.	TX: 17.5 RX: 17.5	15 (raw)	Cable connection
[6.7]	SiGe 0.13	Beamforming 16 Elements	TX: 6200	TX: 43.8	5.6	N.A.

Table 6. I Comparison of 60 GHz wireless communication systems.

6.5. Summary

This chapter presented the most important aspects of TX integration. Measures taken to improve chip ground connection were explained. The effect of poor ground connection was observed in the measurement. A fully differential TX is less sensitive to poor ground connection.

The disruptive PA effect on PLL locking was observed. The problem with solved by separating PLL ground from the ground of the rest of the chip, which includes the PA.

PLL signal phase noise and image-rejection were measured. TX version II has measured Psat of 20 dBm, and P1dB of 17 dBm at 60 GHz.

TX version III was used to for data transmission with data rate of 3.6 Gbit/s (with coding 4.8 Gbit/s) over 15 meters. This is the best result in the class of 60 GHz AFEs without beamforming.

Chapter 7
Conclusion

The large unlicensed 9 GHz band around 60 GHz is dedicated for wireless systems with data rates of several Gbit/s. The high data rate, high security of data transmission, and high frequency reusability make 60 GHz band attractive for many WLAN and WPAN application scenarios. High free-space loss at 60 GHz and relatively low speed of HBT and CMOS transistors for this frequency range make the design of a 60 GHz AFE a challenge.

The goal set in this thesis is the analysis of the challenges and finding solutions for the design of mm-wave transceivers. The work presented in this thesis was focused on design of TX components, which are critical for the performance of the AFE. PLL phase noise was optimized, image-rejection filter and high P1dB PA designed.

The PLL chapter presented basic PLL theory, and different PLL topologies that can be used: third order, forth order PLL and dual loop PLL. Recipes for calculation of PLL parameters were presented, including new optimized recipe for calculating PLL parameters of a forth order PLL. It was shown that by choosing $f_c/f_{PM} = 2.2$ (ratio of crossover frequency and maximum phase margin frequency) we can design PLL with better spur performance, lower LPF noise, smaller charge pump current or smaller capacitor size (for area reduction of integrated PLLs). Using this approach we can reduce spur by up to 10 dB.

The image-rejection filter chapter analyzes the challenges related to the design of the integrated image–rejection filter. The previously published work on 60 GHz filters dealt with on-board filters, where the achievable Q-factor is much higher than for the integrated filters. The analysis presented here is the first on integrated filters for 60 GHz range.

The main problems related to the design of integrated filters arise from the low quality factor of the integrated resonators. The effects are high insertion loss and low selectivity. Two measures to reduce the insertion loss of the image–rejection filters were suggested. One is to design the filter as broadband. The lowest transmission pole should match the signal frequency. This measure deteriorates selectivity, so the minimum required image–rejection will limit the width of the passband. The second measure is to design the filter as broadband with most transmission zeros located at the passband of interest. This measure will improve both the insertion loss and the image–rejection.

In the case when the image is far away from the signal (for IF 12 GHz) and when the filter requirement are not strict, the best option is to design a lumped element integrated filter. Lumped element filters are very small and can be optimized for very good insertion loss and return loss.

The PA chapter presents the basic theory relevant for the design of a high P1dB PA. Power combining techniques are presented and their limits analyzed. The procedure of the PA design was presented. The differential cascode PA topology was chosen with class A operation and power combining of paralleling transistors was implemented. PA simulations for the maximum P1dB were presented.

Difficulties related to crating a mm-wave layout from schematic were discussed. The layout is drawn and optimized in an iterative design procedure and post layout simulation was done. The PA features a symmetrical layout, which inherently gives ideal ac ground on the symmetry axis. This is very important for an wire-bonded chip working on high frequencies, where ground connection via bond wires is poor.

The measured PA had the highest reported P1dB of 17 dBm in the class of SiGe Pas at 60 GHz when it was published.

This TX integration chapter presented the most important aspects of chip integration. Measures taken to improve chip ground connection were explained. The effect of poor ground connection was observed in the measurement. The measurements have confirmed that a fully differential TX is less sensitive to poor ground connection.

PLL signal phase noise and image-rejection were measured. TX version II has measured P_{sat} of 20 dBm, and P1dB of 17 dBm at 60 GHz.

TX version III was used to for data transmission with data rate of 3.6 Gbit/s (with coding 4.8 Gbit/s) over 15 meters. This is the best result in the class of 60 GHz AFEs without beamforming. This result is a proof of a good functionality of the TX chip, and a proof that the solutions used in the TX design are effective.

The future work will focus on the beamforming that will enable NLOS operation when the LOS is blocked. It is a challenging work that should result in a system that is capable of finding the optimum beam direction for the TX and RX with or without the LOS.

References

[1.1] P.F.M. Smulders, "60 GHz radio: prospects and future direction", *Proc. IEEE Benelux Chapter, 10th Symposium on Communications and Vehicular Technology*, Eindhoven (The Netherlands), pp. 1-8. Nov. 2003.

[1.2] http://wireless.fcc.gov/outreach/2004broadbandforum/comments/YDI_benefits60G Hz.pdf Report to FCC, 2002.

[1.3] http://www.continental-wireless-solutions.com/Products/Wireless/PMP/TeraBeam/ Performance60_Ghz.pdf

[1.4] S. Reynolds, B. Floyd, U. Pfeiffer, and T. Zwick, "60GHz transceiver circuits in SiGe bipolar technology," *IEEE ISSCC Dig. Tech. Papers*, Feb. 2004, pp. 442-538.

[1.5] W. Winkler, *et al.*, "60 GHz Transceiver Circuits in SiGe:C BiCMOS Technology," *ESSCIRC*, pp. 83-86, Sept. 2004.

[1.6] http://www.wigwam-project.de/

[1.7] http://www.easy-a.de/index_en.html

[1.8] http://www.ecma-international.org/activities/Communications/ga-2007-015.pdf

[1.9] http://ewh.ieee.org/r6/phoenix/wad/Handouts/bosco.ppt

[1.10] B. Heinemann, H. Rücker, R. Barth, J. Bauer, D. Bolze, E. Bugiel, J. Drews, K.-E. Ehwald, T. Grabolla, U. Haak, W. Höppner, D. Knoll, D. Krüger, B. Kuck, M. Kurps, R. I Marchmeyer, H. H. Richer, P. Schley, D. Schmidt, R. Scholz, B. Tillak, D. W. H.-E. Winkler, W. Wolansky, Y. Yamamoto, and P. Zaumseil, "Novel Collector Design for High-Speed SiGe:C HBTs," IEDM Tech. Dig. 2002, pp. 775-778.

[1.11] J.-S. Rieh, B. Jagannathan, H. Chen, K. Schonenberg, S.-J. Jeng, M. Khater, D. Ahlgren, G. Freeman, S. Subbanna, "Performance and design considerations for high speed SiGe HBT's of f_T/f_{max} =375 GHz / 210 GHz," in *Proc. IEEE Conf. Indium Phosphide and Related Materials*, May 2003, pp. 374-377.

[1.12] B. Floyd, S. Reynolds, U. Pfeiffer, T. Beukema, J. Grzyb, and C. Haymes, "A Silicon 60 GHz Receiver and Transmitter Chipset for Broadband Communications," *ISSCC Dig. Tech. Papers*, pp. 220-221, Feb., 2006.

[1.13] Y. Sun, S. Glisic, F. Herzel, K. Schmalz, E. Grass, W. Winkler and R. Kraemer, "An Integrated 60 GHz Transceiver Front-End for OFDM in SiGe: BiCMOS," in *Proc .of the 16th. Wireless World Research Forum,,* 2006, pp. 121–127.

[2.1] Y. Sun, J. Borngräber, F. Herzel, W. Winkler, "A fully integrated 60 GHz LNA in SiGe:C BiCMOS technology," *BCTM*, Santa Barbara, Oct. 2005, pp. 14-17.

[2.2] K. Schmalz, F. Herzel and M. Piz, "An Integrated 5 GHz Wideband Quadrature Modem in SiGe:C BiCMOS Technology," *36th European Microwave Conference*, Manchester, Sept. 2006. Pp. 1656 – 1659.

[3.1] F.M. Gardner, *Phaselock Techniques*, 2nd ed., New York: John Wiley, 1981.

[3.2] R.E. Best, *Phase-Locked Loops*, 2nd ed., New York: McGraw-Hill, 1993.

[3.3] C.A. Sharpe, "A 3-State Phase Detector Can Improve Your Next PLL Design," *EDN*, pp. 55–59, September 20, 1976.

[3.4] D.R. Stephens, *Phase-Locked Loops for Wireless Applications*, Dordrecht, Kluwer Academic Publisher, 1998.

[3.5] R.C. Dorf, *Modern Control Systems*, 3rd ed., Reading, MA, Addison-Wesley Publishing Company, 1980.

[3.6] F.M. Gardner, "Charge-Pump Phase-Locked Loops," *IEEE Trans. Comm.*, Vol. COM–28, pp. 1849–1858, November 1980.

[3.7] H. R. Rategh, H. Samavati and T. H. Lee, "A CMOS Frequency Synthesizer with an Injection-Locked Frequency Divider for a 5-GHz Wireless LAN Receiver," *IEEE Journal of Solid State Circuits*, vol. 35, No. 5, pp. 780–787, May 2000.

[3.8] National Semiconductor Application Not1 1001, "An Analysis and Performance Evaluation of a Passive Filter Design Technique for Charge Pump PLLs," July. 2001.

[3.9] C. Vaucher, "An Adaptive PLL Tuning System Architecture Combining High Spectral Purity and Fast Settling Time," IEEE J. Solid-State Circuits, vol. 35, pp. 490-502, Apr. 2000.

[3.10] F. Herzel, G. Fischer, H. Gustat, "An Integrated CMOS RF Synthesizer for 802.11a Wireless LAN," IEEE J. Solid-State Circuits, vol. 38, pp. 1767–1770, Oct. 2003.

[3.11] S. Glisic and W. Winkler, "A Broadband Low Spur Fully Integrated BiCMOS PLL for 60 GHz Wireless Applications," *RWS 2006*, San Diego, USA, pp. 451–455.

[3.12] W. Winkler et al., "A Fully Integrated BiCMOS PLL for 60 GHz Wireless Applications," ISSCC Dig. Tech. Papers, Feb. 2005, pp. 406-407.

[3.13] F. Herzel, S. Glisic, S.A. Osmany, C. Scheytt, K. Schmalz, W. Winkler and M. Engels, "A Fully Integrated 48-GHz Low-Noise PLL with a Constant Loop Bandwidth," *SiRF 2008*, Orlando, USA, pp. 82-85.

[4.1] R. Saal and E. Ulbrich, "On the design of filters by synthesis," *IRE Trans.*, *CT-5*, 284–327, Dec. 1958.

[4.2] G. Mattaei, L. Young and E. M. T. Jones, *Microwave Filters, Impedance-Matching Networks, and Coupling Structures*, Artech House, Norwood, MA, 1980.

[4.3] P. I. Richards, "Resistor-transmission-line circuits," *Proc IRE.*, *36*, 217–220, Feb. 1948.

[4.4] J.-A. G. Hong and M. J. Lancaster, *Microstrip Filters for RF/Microwave Applications*, Wiley, New York, 2001.

[4.5] T. Edwards, *Foundations for Microstrip Circuit Design*, Second Edition, Wiley, Chichester, U.K., 1991.

[4.6] M. Kirschning, R. H. Jansen and N. H. L. Koster, "Accurate model for open end effect of microstrip lines," *Electronics Letters*, *17*, 123–125, Feb. 1981.

[4.7] K. C. Gurpta, R. Garg, I. Bahl and P. Bhartis, *Microstrip Lines and Slotlines*, Second Edition, Artech House, Boston, 1996.

[4.8] R. Garg and I. J. Bahl, "Characterization of coupled microstriplines," *IEEE Trans.*, *MTT-27*, July 1979, 700-705. Correction in *IEEE Trans.*, *MTT-28*, March 1980, p. 272.

[4.9] M. Kirschning, R. H. Jansen, "Accurate wide-range design equations for parallel coupled microstrip lines," *IEEE Trans.*, *MTT-32*, 83–90, Jan. 1984. Corrections in *IEEE Trans.*, *MTT-33*, p. 228, March 1985.

[4.10] R. A. Pucel, D. J. Masse and C. P. Hartwig, "Losses in microstrip," *IEEE Trans.*, *MTT-16*, 342–350, June 1968. Corrections in *IEEE Trans.*, *MTT-16*,p. 1064, Dec. 1968.

[4.11] IEEE 802.15 WPAN Millimeter Wave Alternative PHY Task Group 3c [Online]. http://ieee802.org/15/pub/TG3c_contributions.html

[4.12] I. Ferrer and J. Sverin, "A 60 GHz Image Rejection Filter Manufactured Using a High Resolution LTCC Screen Printing Process," in Proc. EUMC, 2003, pp. 423–425.

[4.13] D.S. Jun, D.Y. Lee, D-Y. Kim, S.S. Lee and E.S. Nam, "A Narrow Bandwidth Band-Pass Filter with Symmetrical Frequency Characteristics," ETRI Journal, vol. 27, pp. 643–646, Oct. 2005.

[4.14] S. S. Choi and D. C. Park, "60-GHz Band Dual-Mode Microstrip Ring Resonator Bandpass Filter Using Micromachining Technology," International Journal of Infrared and Millimeter Waves, vol. 29, pp. 961–967, Nov. 2007.

[4.15] T. Shimizu and T. Yoneyama, "60 GHz Bandpass Filter Using NRD Guide E-Plane Resonator," IEICE Transactions on Electronics, vol. E89–C, num. 12, pp. 1851–1857, Dec. 2006.

[4.16] E. van der Hejden, M. Notten, G. Dolmans, H. Veenstra and R. Pijper, "On-chip third-order band-pass filters for 24 and 77 GHz car radar," in International Microwave Symposium Tech. Dig., pp. 697–700, San Francisco, Jun. 2006.

[4.17] B. Dehlink, M. Engl, K. Aufinger and H. Knapp, "Integrated Bandpass Filter at 77 GHz in SiGe Technology," IEEE Microwave and Wireless Components Letters, vol. 17, May 2007.

[4.18] S. Glisic and C. Scheytt, "Integrated Compact Microstrip Filters for 60 GHz Applications," in Proc.38 IEEE EuMC, Oct. 2008.

[5.1] G. Gonzalez, "Microwave Transistor Amplifiers Analysis and Design," *Practice Hall*, 1997.

[5.2] B. Razavi, "RF Microelectronics," *Pearson Education, Inc*. 2003.

[5.3] S.C. Cripps, "RF Power Amplifiers for Wireless Communications," *Artech House, Inc.* 2006.

[5.4] B. Floyd, S. Reynolds, U. Pfeiffer, T. Beukema, J. Gryzb, C. Haymes, "A Silicon 60GHz Receiver and Transmitter Chipset for Broadband Communications," in Proc. ISSCC, 2006, pp. 649–658

[5.5] Y. Sun, S. Glisic, F. Herzel, K. Schmalz, E. Grass, W. Winkler and R. Kraemer, "An Integrated 60 GHz Transceiver Front-End for OFDM in SiGe: BiCMOS," in Proc .of the 16th. Wireless World Research Forum,, 2006, pp. 121–127.

[5.6] U. Pfeiffer, "A 20dBm Fully-Integrated 60GHz SiGe Power Amplifier with Automatic Level Control," in Proc. ESSCIRC, pp. 356–359, Sept. 2006.

[5.7] C-H. Wang, Y-H. Cho, C-S. Lin, H. Wang, C-H. Chen, D-C. Niu, J. Yeh, C-Y. Lee and J. Chern, "A 60 GHz Transmitter with Integrated Antenna in 0.18μm SiGe BiCMOS Technology," *IEEE International Solid-State Circuits Conference*, pp.186–187, Feb. 2006.

[5.8] U. Pfeiffer, D. Goren, B. A. Floyd and S. K. Reynolds, "SiGe Transformer Matched Power Amplifier for Operation at Millimeter-Wave Frequencies," *European Solid-State Circuits Conference*, vol. 29, pp. 961–967, Nov. 2007.

[5.9] H. Li, H-M. Rein, T. Suttorp and J. Boeck, "Fully Integrated SiGe VCOs with Powerful Output Buffer for 77-GHz Automotive Radar Systems and Applications around 100 GHz," *IEEE Journal of Solid-State Circuits,* vol. 39, no. 10, pp. 1650–1658, Oct. 2004.

[5.10] A. Komijani and A. Hajimiri, "A Wideband 77GHz, 17.5dBm Power Amplifier in Silicon," *Custom Integrated Circuits Conference,* pp. 566–569, Sept. 2005.

[5.11] U. Pfeiffer, S. Reynolds and B. Floyd, "A 77 GHz SiGe Power Amplifier for Potential Applications in Automotive Radar Systems," *Radio Frequency Integrated Circuits Symposium*, pp. 91–94, June 2004.

[5.12] A. Valdes-Garcia, S. Reynolds and U. Pfeiffer, "A 60 GHz Class-E Power Amplifier in SiGe," in *Proc. ASSCC,*, pp. 199–202, Nov 2006.

[5.13] E. Afshari, H. Bhat, X. Li and A. Hajimiri, "Electrical Funnel: A Broadband Signal Combining Method," *IEEE International Solid-State Circuits Conference*, pp.206–207, Feb. 2006.

[5.14] D. Chowdhury, P. Reynaert and A. M. Hajimiri, "A 60 GHz 1V +12.3dBm Transformer-Coupled Wideband PA in 90nm CMOS," *IEEE International Solid-State Circuits Conference*, pp.560–561, Feb. 2008.

[5.15] T. Suzuki, Y. Kawano, M. Sato, T. Hirose and K. Joshin, "60 and 77 GHz Amplifiers in Standard 90nm CMOS," *IEEE International Solid-State Circuits Conference*, pp.562–563, Feb. 2008.

[5.16] Y. Jin, M. A. T. Sanduleanu, E. A. Rivero and J. R. Long, "A Millimeter-Wave Power Amplifier with 25dB Power Gain and +8dBm Saturated Output Power" in *Proc. ESSCIRC,* pp. 276–279, Sept. 2007.

[5.17] S. Glisic and J.C. Scheytt, "A 13.5-to-17 dBm P1dB, Selective High-Gain Power Amplifier for 60 GHz Applications in SiGe" *Proc. BCTM* 2008.

[6.1] S. Glisic, Y. Sun, F. Herzel, W. Winkler, M. Piz, E. Grass and C. Scheytt, "A Fully Integrated 60 GHz Transmitter Front-End with a PLL, an Image-rejection Filter and a PA in SiGe," *ESSCIRC*, Edinburgh, UK, Sept. 2008, pp. 242-245.

[6.2] E. Grass, I. Siaud, S. Glisic, M. Ehrig, Y. Sun, J. Lehman, M. H. Hamon, A.M. Ulmer-Moll, P. Pagani, R. Kraemer and C. Scheytt, "Asymmetric Dual-Band UWB / 60GHz Demonstrator," *PIMRC*, Cannes, France, Sept., 2008, pp. 1-6.

[6.3] S. Reynolds, A. Valdes-Garcia, B. Floyd, T. Beukema, B. Gaucher, D. Liu, N. Hoivik and B. Orner, "Second Generation 60-GHz Transceiver Chipset Supporting Multiple Modulations at Gb/s data rates", *BCTM*, Sept. 2007, pp. 192-197.

[6.4] A. Tomkins, R.A. Aroca, T. Yamamoto, S.T. Nicolson, Y. Doi, S.P. Voinigescu, "A Zero-IF 60GHz Transceiver in 65nm CMOS with > 3.5Gb/s Links," *CICC*, San Jose, CA, Sept. 2008, pp. 471-474.

[6.5] C. Marcu, D. Chowdhury, C. Thakkar, J. Park, L. Kong, M. Tabesh, Y. Wang, B. Afshar, A. Gupta, A. Arbabian, S. Gambini, R. Zamani, E. Alon, A. M. Niknejad, "A 90nm CMOS Low-Power 60 GHz Transceiver With Integrated Baseband Circuitry," *IEEE Journal of Solid-State Circuits*, vol. 44, Dec. 2009.

[6.6]   S. Pinel, P. Sen, S. Sarkar, B. Perumana, D. Dawn, D. Yeh, F. Barale, M. Leung, E. Juntunen, P. Vadivelu, K. Chuang, P. Melet, G. Iyer, and J. Laskar, "60GHz Single-Chip CMOS Digital Radios and Phased Array Solutions for Gaming and Connectivity", *IEEE Journal on Selected Areas of Communications*, vol. 27, No. 8, Oct. 2009, pp. 1347-1357.

[6.7]   A. Valdes-Garcia, S. Nicolson, J.-W. Lai, A. Natarajan, P.-Y. Chen, S. Reynolds, J.-H. Conan Zhan, B. Floyd' "A SiGe BiCMOS 16-Element Phased-Array Transmitter for 60GHz Communications" *ISSCC*, San Francisco, CA, Feb. 2010, pp. 218-219.

List of Figures

Figure 1. 1 Atmospheric absorption of mm-wave frequencies over a 1-km path (from [1.3]) 2

Figure 1. 2 Spectrum allocation in the unlicensed 57-66 GHz band. 3

Figure 1. 3 Some possible applications for 60 GHz radio. 3

Figure 2. 1 AFE block diagram. 9

Figure 2. 2 Transmitter version I block diagram. 10

Figure 2. 3 Fully differential transmitter version II block diagram. 11

Figure 2. 4 Fully integrated differential transmitter version III block diagram. 11

Figure 2. 5 Simplified schematic of the upconversion mixer. 12

Figure 2. 6 Simplified schematic of the preamplifier. 13

Figure 3. 1 Basic type I phase–locked loop. 16

Figure 3. 2 Simple lowpass filter for the type I PLL. 17

Figure 3. 3 Type I PLL signals when the PLL is locked. 18

Figure 3. 4 Type I PLL signals while the PLL is locking. 19

Figure 3. 5 Basic type II phase–locked loop. 20

Figure 3. 6 PFD symbol a). PFD response for $\omega_A > \omega_B$ b), and $\omega_A = \omega_B$ with ω_A lagging c) 21

Figure 3. 7 PFD state diagram. 21

Figure 3. 8 PFD implementation. 21

Figure 3. 9 Ideal charge pump model. 22

Figure 3. 10 Type II PLL linear model. 23

Figure 3. 11 a) The simplest type II PLL LP filter. b) Type II PLL filter with zero in transfer function. c) Type II PLL filter with the ripple suppression capacitor C_1 (third order PLL). d) Type II PLL filter with the ripple suppression filter stage R_3C_3 (fourth order PLL) 24

Figure 3. 12 Transfer function pole loci for a stable and an unstable system. 26

Figure 3. 13 Open loop transfer function phase margin is defined for unity gain frequency shown in a). The margin is shown in b). 26

Figure 3. 14 Type II PLL transfer function pole loci. 27

Figure 3. 15 PLL input phase transfer function normalized for the crossover frequency, ω_c, for different values of phase margin and division ratio N = 1. 28

Figure 3. 16 PLL VCO phase noise transfer function normalized for the crossover frequency, ω_c, for different values of the PLL phase margin. 29

Figure 3. 17 PLL LPF phase noise transfer function normalized for the crossover frequency, ω_c, for different values of the PLL phase margin. 30

Figure 3. 18 Relative value of Spur for different phase margin values and the same PLL bandwidth, LPF noise level and capacitor C_2. 35

Figure 3. 19 Relative value of Spur for different PLL bandwidth values and the same phase margin and capacitor C_2 and scaled LPF noise level. 36

Figure 3. 20 Relative value of Spur for different values of f_c/f_{PM} ratio and the same phase margin, PLL bandwidth, capacitor C_2 and LPF noise level. 38

Figure 3. 21 Phase plots for PLL parameters in Table 3. III for f_c/f_{PM} = 1 and f_c/f_{PM} = 2.2. 38

Figure 3. 22 Settling behaviour o f PLLs with PLL parameters in Table III for f_c/f_{PM} = 1 (red line) and f_c/f_{PM} = 2.2 (blue line) (a), settling of the voltage at capacitor C_2 (b). 39

Figure 3. 23 Relative spur level dependency on the size of capacitor C_2 for f_c/f_{PM} = 1 and f_c/f_{PM} = 2.2. 40

Figure 3. 24 Topology of a dual-loop PLL. 41

Figure 3. 25 56 GHz PLL micrograph. 44

Figure 3. 26 Close spectrum (20 MHz) of the PLL output signal. 45

Figure 3. 27 Broad spectrum (500 MHz) of the PLL output signal. 46

Figure 3. 28 Phase noise spectrum of a 100 MHz baseband signal. 46

Figure 3. 29 48 GHz PLL micrograph. 47

Figure 4. 1 Two–port network showing network variables. 50

Figure 4. 2 A bandpass filter response with specifications. 52

Figure 4. 3 A lowpass Butterworth filter frequency response (n=1,3,10)(a),and in dB(b). 52

Figure 4. 4 Lowpass filter prototype (a) and its dual (b) for all–pole filters with ladder network structures. 53

Figure 4. 5 Transformation of lowpass prototype filter elements to practical lowpass filter. 56

Figure 4. 6 Transformation of lowpass prototype filter elements to practical bandpass filter. 57

Figure 4. 7 Admittance and impedance inverters. 57

Figure 4. 8 Admittance inverters used to implemented a series inductance with a shunt capacitance. 57

Figure 4. 9 Typical admittance inverters from lumped elements. 58

Figure 4. 10 Admittance inverter from lumped and transmission line elements. 58

Figure 4. 11 Lowpass filter as cascade of J–inverters and shunt capacitors. 59

Figure 4. 12 Bandpass filter as cascade of J–inverters and parallel LC resonators. 60

Figure 4. 13 Bandpass filter as cascade of J–inverters and distributed elements. 61

Figure 4. 14 Richards' transformation of the real angular frequency ω into distributed angular frequency Ω (a), Chebyshev response after Richards' transformation (b). 62

Figure 4. 15 Richards' transformation of inductor (a) and capacitor (b) from p–plane into short– and open–circuited transmission line in the t–plane. 62

Figure 4. 16 Layout of an end–coupled microstrip bandpass filter. 63

Figure 4. 17 Microstrip discontinuities and their equivalent circuits. Open–end (a) and gap (b). 64

Figure 4. 18 Layout of a parallel–coupled microstrip bandpass filter. 65

Figure 4. 19 Layout of a hairpin microstrip bandpass filter. 67

Figure 4. 20 Typical input or output couplings of a coupled resonator filter. Coupled–line coupling a), tapped–line coupling b). 67

Figure 4. 21 The external quality factor parameter extraction. 68

Figure 4. 22 The coupling coefficient parameter extraction. 69

Figure 4. 23 Frequency response of a lowpass 4th order Chebyshev filter (red), with transmission zeros at 1.4 (blue) and 1.8 (pink) of the normalized frequency. 70

Figure 4. 24 An example of a four–pole microstrip quadruplet filter (a), and a three–pole microstrip trisection filter (b). 70

Figure 4. 25 Circuit representations of lossy reactive elements (a) and (b) and series and parallel resonators (c) and (d). 71

Figure 4. 26 Frequency response of a lowpass 5th order Butterworth filter with lumped elements for different quality factor values of the filter elements. 72

Figure 4. 27 Frequency response of a 60 GHz bandpass filter with lumped elements for different quality factor values of the filter elements. 73

Figure 4. 28 Frequency response of a 60 GHz bandpass filter with lumped elements for different bandwidth values. 73

Figure 4. 29 Substrate cross–section (not to be scaled): a) physical, b) for Momentum simulations, c) fitted substrate for measured data. 77

Figure 4. 30 a) 3rd order Chebyshev filter layout. b) Simulated filter S parameters with and without dissipation. 77

Figure 4. 31 a) Trisection filter layout. b) Simulated filter S parameters with and without dissipation. 78

Figure 4. 32 a) Photo of the manufactured filter with the deembeding structures: short, open and thru. b) Simulated filter S parameters. 80

Figure 4. 33 S parameters of the hairpin filter: measured and simulated with fitted substrate 80

Figure 4. 34 Filter layout with ports for Momentum simulation. 81

Figure 4. 35 Simulated and measured S parameters of the filter. 81

Figure 4. 36 Simulated S parameters with ohmic losses and radiation (i.e. with dissipation) and without. 82

Figure 4. 37 Photo of the manufactured filter. 82

Figure 4. 38 Simulated S parameters of the filter with and without pads. 83

Figure 4. 39 Measured and simulated S parameters of the filter with pads. 83

Figure 4. 40 Schematic of the lumped element filter. 84

Figure 4. 41 Filter layout used for Momentum simulation. 84

Figure 4. 42 Measured and simulated S parameters of the lumped filter. 85

Figure 4. 43 Simulated bandpass filter S parameters with and without dissipation. 85

Figure 4. 44 Simulated and measured S parameters of the on–board filter. 86

Figure 5. 1 A block diagram of a one-stage microwave amplifier. 90

Figure 5. 2 a) Stability factor for a HBT npn201_8 transistor in a common emitter configuration. b) The source and load stability circles at 20 GHz ($K \cong 0.6$) with the stability region outside of the circles. The unity circle is $|\Gamma| = 1$ 92

Figure 5. 3 The Smith chart with constant gain G_A and G_p circles for an HBT npn201_8 transistor in IHP H1 technology in a common emitter configuration at 60 GHz. 94

Figure 5. 4 Amplification and distortion of a small and large signal (voltage to current) in a bipolar transistor with common emitter configuration. 95

Figure 5. 5 Typical P_{out} vs. P_{in} curve showing the 1dB compression point and regions of low and high distortion. 96

Figure 5. 6 Typical measurement of the relative phase for a high-linearity PA (class A) and low-linearity PA (class AB). P_{in} is in log scale. 97

Figure 5. 7 Output spectrum of an amplifier around the input tones. 97

Figure 5. 8 Two cascaded amplifiers with equal gain 98

Figure 5. 9 Load lines and quiescent points for different PA classes of operation with the sinusoidal input signal and resulting output waveforms. 100

List of Figures

Figure 5. 10 Load-pull measurement setup...100

Figure 5. 11 Power combining techniques: Parallelizing transistors a) and using hybrid dividers to split input signal and hybrid combiners to add output signals. 102

Figure 5. 12 a) Micrograph of the test structures b) Micrograph of the meandered TL. 103

Figure 5. 13 Comparison of measured and simulated S_{21} parameter of the meandered line in the TM1..104

Figure 5. 14 Simplified ADS schematic used to simulate the optimum output impedance for P1dB power level. Each transistor in the schematic represents four transistors with eight fingers. The encircled resistor R2 is used to model the double power consumption and related self heating in the differential circuit...106

Figure 5. 15 a) The marked points in the Smith chart represent the values of the load impedance for which the P1dB was simulated. b) Power contours representing the load impedances for optimal P1dB (14.52 dBm) and 1,2,3 and 4 dB lower P1dB values. 107

Figure 5. 16 Simplified PA schematic showing one half of the differential PA topology. 108

Figure 5. 17 Comparison of the simulated S-parameters of the output matching network with the output transmission line and the RF pad in the schematic (lumped elements) and layout (distributed elements). 111

Figure 5. 18 Power amplifier layout 112

Figure 5. 19 Structure for ADS Momentum EM simulation. Green is top metal layer; blue is one metal layer lower..112

Figure 5. 20 Power amplifier chip photo. 113

Figure 5. 21 Small-signal S-parameter measurement setup for on-wafer measurement. The DUT is the power amplifier. 114

Figure 5. 22 Large-signal measurement setup for on-wafer measurement. The DUT is the power amplifier. 114

Figure 5. 23 Comparison of measured and simulated S21 and S12 parameters.............................115

Figure 5. 24 Comparison of measured and simulated S11 and S22 parameters.............................115

Figure 5. 25 PA output power and gain vs. input power at 61.5 GHz for 800 mW power consumption. 116

Figure 5. 26 PA output power and gain vs. input power at 65 GHz for 800 mW power consumption. 116

Figure 5. 27 PA power-added efficiency at 61.5 GHz for 600 mW power consumption. Maximum PAE of 10.2 % is reached for input power of –3 dBm. 117

Figure 5. 28 Measured AM to PM distortion of the PA. 119

Figure 5. 29 Saturated output power at the output of the TX (PA) versus IF input frequency from 1 to 10 GHz. The PLL up-conversion frequency is 56 GHz resulting in the output frequency range of 57 to 66 GHz. 119

Figure 5. 30 Temperature measurement on the TX board done with an IC camera showing chip temperature of 60,6°C. 120

Figure 6. 1 Close–up photo of a bonded TX chip. 124

Figure 6. 2 Version I transmitter chip photo. 125

Figure 6. 3 Transmitter board photo with chip close-up. 126

Figure 6. 4 Analog front-end measurement setup. 126

Figure 6. 5 Version I AFE measurement setup photo. 127

Figure 6. 6 Constellation diagram for QPSK, ¾ coding, OFDM signal with 360 Mbit/s data rate over 5 m distance. 127

Figure 6. 7 Fully differential version I transmitter chip photo 128

Figure 6. 8 Block diagram of the asymmetric dual-band demonstrator. 128

Figure 6. 9 Maximum and minimum distance of communication for different data rates. 129

Figure 6. 10 Version III fully integrated transmitter chip photo 130

Figure 6. 11 Fully mounted single-chip TX and RX board. 130

Figure 6. 12 Version III AFE measurement setup photo. 131

Figure 6. 13 Constellation diagram for 16QAM, ¾ coding, OFDM signal with 3.6 Gbit/s data rate over 5 m distance. 131

Figure 6. 14 Constellation diagram for QPSK, 2/3 coding, OFDM signal with 1.6 Gbit/s data rate over 15 m distance. 132

Figure 6. 15 Constellation diagram for 16QAM, ¾ coding, OFDM signal with 3.6 Gbit/s data rate over 15 m distance. 132

List of Tables

Table 1. I Overview of IHP's SiGe:C BiCMOS technology. 4

Table 3. I Calculated PLL Parameters for different phase margins. 34

Table 3. II Calculated PLL Parameters for different PLL bandwidths 35

Table 3. III Calculated PLL Parameters using old and new technique for different ratios of PLL bandwidth and maximum phase margin frequencies 37

Table 3. IV Calculated PLL Parameters using old and new technique for minimum spur, C_2 capacitor and LPF phase noise. 38

Table 3. V Calculated PLL Parameters using old and new technique for different ratios of PLL bandwidth and maximum phase margin frequencies. 40

Table 3. VI Calculated PLL Parameters for Different PLL Topologies. 43

Table 4. I Comparison of 60 and 77 GHz Filters. 76

Table 5. I Degradation of output P1dB due to amplifier cascading for two and three stages 98

Table 5. II Measured PA performance for 800 mW power consumption. 117

Table 5. III Measured PA performance for 600 mW power consumption. 118

Table 5. IV Comparison of mm-wave power amplifiers. 121

Table 6. I Comparison of 60 GHz wireless communication systems. 134

Publications

Journals

1. **S. Glisic**, J.C. Scheytt, Y. Sun, F. Herzel, R. Wang, K. Schmalz, M. Elkhouly, Ch.-S. Choi *Fully-Integrated 60 GHz Transceiver in SiGe BiCMOS, RF Modules, and 3.6 Gbit/s OFDM Data Transmission* International Journal of Microwave and Wireless Technologies, 3(2), 139–145 (2011)

2. F. Herzel, **S. Glisic**, W. Winkler *Integrated Frequency Synthesizer in SiGe BiCMOS Technology for 60 GHz and 24 GHz Wireless Applications* Electronics Letters 43(3), 154 (2007)

3. L. Wang, **S. Glisic**, J. Borngräber, W. Winkler, J.C. Scheytt *A Single-Ended Fully Integrated 77/79 GHz Radar Receiver in SiGe Technology* IEEE Journal of Solid State Circuits 43(9), 1897 (2008)

4. M. Elkhouly, Ch.-S. Choi, **S. Glisic**, F. Ellinger, J.C. Scheytt *A 60 GHz Eight-Element Phased-Array Receiver with Interference Mitigation in 0.25 μm SiGe BiCMOS Technology* IEEE Transactions on Microwave Theory and Technique (submitted)

Conferences

1. **S. Glisic**, W. Winkler *A Broadband Low Spur Fully Integrated BiCMOS PLL for 60 GHz Wireless Applications* Proc. IEEE Radio and Wireless Symposium – RWS '06, 451

2. **S. Glisic**, L. Wang *SiGe ICs for the 77 GHz Automotive Radar* Proc. EEEfCOM Workshop Hochfrequenztechnik, Komponenten, Module und EMV (2006)

3. J.C. Scheytt, Y. Sun, **S. Glisic**, F. Herzel, K. Schmalz, E. Grass, W. Winkler, R. Kraemer *60 GHz SiGe Transceiver Frontend-Ics für die drahtlose Nahfeldkommunikation* Proc. EEEfCOM Workshop Hochfrequenztechnik, Komponenten, Module und EMV (2006)

4. Y Sun, **S. Glisic**, F. Herzel *A Fully Differential 60 GHz Receiver Front-End with Integrated PLL in SiGe:C BiCMOS* Proc. European Microwave Integrated Circuits Conference, 198 (2006)

5. Y. Sun, **S. Glisic**, M. Piz, F. Herzel, K. Schmalz, E. Grass, W. Winkler, J.C. Scheytt, R. Kraemer *An Integrated 60 GHz Transceiver Front-End for OFDM in SiGe BiCMOS* Digest of the 4th Joint Symposium on Opto- and Microelectronic Devices and Circuits (SODC '06), 93

6. E. Grass, M. Piz, F. Herzel, K. Schmalz, Y. Sun, **S. Glisic**, M. Krstic, K. Tittelbach-Helmrich, M. Ehrig, W. Winkler, R, Kramer, J.C. Scheytt *Broadband Wireless Communication at 60 GHz: Systems, Circuits and Technologies* Workshop " From Research to Innovation", Szczecin, May 17-19, 2006, Poland

7. R. Kraemer, Y. Sun, **S. Glisic**, M. Piz, F. Herzel, K. Schmalz, E. Grass, W. Winkler, J.C. Scheytt *An Integrated 60 GHz Transceiver Front-End for OFDM in SiGe: BiCMOS* 4th Joint Symposium on Opto- and Microelectronic Devices and Circuits (SODC 2006), Duisburg, September 03 -08, 2006, Germany

8. Y Sun, **S. Glisic**, F. Herzel *A Fully Differential 60 GHz Receiver Front-End with Integrated PLL in SiGe:C BiCMOS* 5th Workshop High-Performance SiGe BiCMOS for Wireless and Broadband Communication, Frankfurt (Oder), September 25-26, 2006, Germany

9. Y. Sun, **S. Glisic**, F. Herzel, K. Schmalz, E. Grass, W. Winkler, R. Kraemer *An Integrated 60 GHz Transceiver Front-End for OFDM in SiGe BiCMOS* WWRF 16, Shanghai, April 26-28, 2006, China

10. S. Chartier, B. Schleicher, F. Korndörfer, **S. Glisic**, G.G. Fischer, H. Schumacher *A Fully Integrated Fully Differential Low-Noise Amplifier for Short Range Automotive Radar Using a SiGe:C BiCMOS Technology* Proc. European Microwave Week, 407 (2007)

11. S. Chartier, L. Liu, G.G. Fischer, S. Glisic, H. Höhnemann, A. Trasser, H. Schumacher *SiGe Millimeter-Wave Dynamic Frequency Divider with Enhanced Sensitivity Incorporating a Transimpedance Stage* Proc. European Microwave Week, 84 (2007)

12. G.G. Fischer, **S. Glisic** *A SiGe:C BiCMOS Technology for 77-81 GHz Automotive Radar Applications* Proc. European Microwave Week, WSW5, (Automotive High Frequency Electronics – KOKON) (2007)

13. E. Grass, F. Herzel, M. Piz, K. Schmalz, Y. Sun, **S. Glisic**, M. Krstic, K. Tittelbach-Helmrich, M. Ehrig, W. Winkler, J.C. Scheytt, R. Kraemer *60 GHz SiGe-BiCMOS Radio for OFDM Transmission* Proc. ISCAS 2007, 1979

14. L. Wang, **S. Glisic**, J. Borngräber, W. Winkler, J.C. Scheytt *A Single-Ended 79 GHz Radar Receiver in SiGe Technology* Proc. BCTM, 14.4 (2007)

15. F. Herzel, **S. Glisic**, K. Schmalz, Y. Sun, J.C. Scheytt *60 GHz Analog Frontend Circuits in SiGe BiCMOS Technology* EEEfCOM 2007, Ulm, June 21, 2007, Germany

16. Ch.-S. Choi, E. Grass, F. Herzel, M. Piz, K. Schmalz, Y. Sun,. **S. Glisic**, M. Krstic, K. Tittelbach-Helmrich, M. Ehrig, W. Winkler, R. Kraemer, J.C. Scheytt *60 GHz OFDM System Demonstrators in SiGe BiCMOS: State-of-the-Art and Future Development* Proc. PIMRC 2008

17. G.G. Fischer, **S. Glisic** *SiGe:C BiCMOS Technologies for Automotive Radar Applications* SiGe, Ge, and Related Compounds3: Materials, Processing, and Devices, Pennington : The Electrochemical Society, ECS Transactions 16(10), 1041 (2008)

18. G.G. Fischer, **S. Glisic** *Temperature Stability and Reliability Aspects of 77 GHz Voltage Controlled Oscillators in a SiGe:C BiCMOS Technology* Proc. SiRF 2008, 171

19. **S. Glisic**, J.C. Scheytt *A 13.5-to-17 dBm P1dB, Selective High-Gain Power Amplifier for 60 GHz Applications in SiGe* Proc. BCTM 2008, 65

20. **S. Glisic**, Y. Sun, F. Herzel, W. Winkler, M. Piz, E. Grass, J.C. Scheytt *A Fully Integrated 60 GHz Transmitter Front-End with PLL, an Image-Rejection Filter and a PA in SiGe* Proc. IEEE European Solid-State Circuits Conference (ESSCIRC), 242 (2008)

21. **S. Glisic**, J.C. Scheytt *An Integrated and On-board Compact Microstrip Filters for 60 GHz Applications* Proc. EUMW – European Microwave Week 2008, 1386

22. E. Grass, I. Siaud, **S. Glisic**, M. Ehrig, Y. Sun, J. Lehmann, M.H. Hamon, A.M. Ulmer-Moll, P. Pagani, R. Kraemer, J.C. Scheytt *Asymmetric Dual-Band UWB / 60 GHz Demonstrator* Proc. PIMRC 2008

23. F. Herzel, **S. Glisic**, S. Osmany, J.C. Scheytt, K. Schmalz, M. Engels *A Fully Integrated 48-GHz-Low-Noise PLL with a Constant Loop Bandwidth* Proc. SiRF 2008, 82

24. Y. Sun, **S. Glisic** *60 GHz Front-End Design in SiGe BiCMOS Technology* Proc. Asia Pacific Microwave Conference 2008, B7-07

25. W. Winkler, **S. Glisic**, E. Grass, F. Herzel, J.C. Scheytt, L. Wang *Millimeter-Wave Circuits in SiGe BiCMOS Technology for Radar and Communication* MWE 2008, Microwave Workshop & Exhibition, Yokohama, November 26-28, 2008, Japan

26. J.C. Scheytt, **S. Glisic**, Y. Sun, C.S. Choi, M. Elkhouly, F. Herzel, E. Grass *60 GHz OFDM Transceiver RF Frontend Design in SiGe BiCMOS* Proc. IEEE Radio & Wireless Symposium 2010

27. J.C. Scheytt, **S. Glisic**, P. Ostrovskyy, H. Gustat, K. Schmalz, J. Borngräber, S.A. Osmany, F. Herzel, B. Heinemann, H. Rücker, D. Knoll, B. Tillack *SiGe BiCMOS Circuits for High-Frequency Communications and Sensing Applications* Proc. SiRF 2010

28. M. Elkhouly, **S. Glisic**, J.C. Scheytt *A High Output P1dB 60-GHz up-Conversation Image Rejection Mixer in 0.25 µm SiGe Technology* Silicon Monolithic Integrated Circuits in RF Systems, SiRF 2010, New Orleans, January 11 – 13, 2010, USA

29. J.C. Scheytt, **S. Glisic**, Y. Sun, K. Schmalz, W. Winkler, W. Debski, F. Herzel *mm-Wave Transceiver and Component Design for 60, 94 and 122 GHz in SiGe BiCMOS Technology* 6th Joint Symposium on Opto- & Micro-electronic Devices and Circuits (SODC 2010), Berlin, October 03 – 10, 2010, Germany

List of Acronyms and Symbols

ac	Alternating current
ADC	Analog-to-Digital Converter
ADS	Advanced Design System
AFE	Analog Front-End
AGC	Automatic Gain Control
AWG	Arbitrary Waveform Generator
B	Susceptance, Bandwidth
BB	Baseband
BER	Bit Error Rate
BiCMOS	Bipolar Complementary Metal-Oxide-Semiconductor
BPF	Band-pass Filter
BPSK	Binary Phase-shift Keying
BV_{CE0}	Collector-emitter breakdown voltage
BV_{CB0}	Collector-base breakdown voltage
BW	Bandwidth
C	Capacitor
CMOS	Complementary Metal Oxide Semiconductor
CPE	Common Phase Error
DAC	Digital-to-Analog Converter
dB	Decibel
dc	Direct current

DFF	D Flip-Flop
DUT	Device Under Test
EM	Electromagnetic
FER	Frame Error Rate
FM	Frequency Modulation
f_{max}	Maximum frequency of oscillation
f_T	Transit frequency
G	Conductance
GHz	Gigahertz
GSG	Ground-Signal-Ground
HBT	Heterojunction Bipolar Transistor
HDTV	High Definition Television
I	Current
IC	Integrated Circuit
IEEE	Institute of Electrical and Electronics Engineers
IF	Intermediate Frequency
IM	Intermodulation
ISM	Industrial Scientific Medical
L	Inductor
LC	Inductor-Capacitor
LNA	Low-noise Amplifier
LPF	Low-pass filter
LO	Local Oscillator
LOS	Line-of-Sight

mm-wave	Millimetre wave
MIM	Metal-Insulator-Metal
MHz	Megahertz
MMIC	Monolithic Microwave Integrated Circuit
MOS	Metal-Oxide-Semiconductor
NLOS	Non Line-of-Sight
OFDM	Orthogonal Frequency Division Multiplex
P1dB	1dB compression point
PA	Power Amplifier
PCB	Printed Circuit Board
PD	Phase Detector
P$_{DC}$	Power dissipation
PFD	Phase-frequency Detector
PLL	Phase-locked Loop
P$_{sat}$	Saturated output power
QAM	Quadrature amplitude modulation
QPSK	Quadrature Phase-shift Keying
R	Resistor, Resistance
RF	Radio Frequency
RMS	Root Mean Square
RX	Receiver
SiGe	Silicon-Germanium
SiGe:C	Silicon-Germanium Carbon
SNR	Signal-to-noise Ratio

S-parameters	Scattering parameters
SPI	Serial Peripheral Interface
TL	Transmission Line
TX	Transmitter
UWB	Ultra Wide Band
V	Voltage
VBIC	Vertical Bipolar Intercompany Model
VCO	Voltage-Controlled Oscillator
X	Reactance
Y	Admittance
Z	Impedance
ZIF	Zero Intermediate Frequency

Die VDM Verlagsservicegesellschaft sucht für wissenschaftliche Verlage abgeschlossene und herausragende

Dissertationen, Habilitationen, Diplomarbeiten, Master Theses, Magisterarbeiten usw.

für die kostenlose Publikation als Fachbuch.

Sie verfügen über eine Arbeit, die hohen inhaltlichen und formalen Ansprüchen genügt, und haben Interesse an einer honorarvergüteten Publikation?

Dann senden Sie bitte erste Informationen über sich und Ihre Arbeit per Email an *info@vdm-vsg.de*.

Sie erhalten kurzfristig unser Feedback!

VDM Verlagsservicegesellschaft mbH
Dudweiler Landstr. 99　　　　　　　Telefon +49 681 3720 174
D - 66123 Saarbrücken　　　　　　　Fax　　　 +49 681 3720 1749
www.vdm-vsg.de

Die VDM Verlagsservicegesellschaft mbH vertritt

Printed by Books on Demand GmbH, Norderstedt / Germany